4th Euro-Mediterranean Conference on Materials and Renewable Energies (EMCMRE-4)

Journal of Physics: Conference Series Volume 1081

Marrakech, Morocco
8-11 May 2017

Editors:

Hamid Oughaddou
Abdelkader Kara
Azzedine Bendounan
Karima Lasri
Andrew Mayne
Abedlilah Kaddouri
Nabil Rochdi
Abdelaziz El Boujlaidi
Jean-Louis Lemaire
Khalil Amine
Laurene Tetard
Ismael Saadoune

ISBN: 978-1-5108-7089-5
ISSN: 1742-6588

Printed from e-media with permission by:

Curran Associates, Inc.
57 Morehouse Lane
Red Hook, NY 12571

CURRAN ASSOCIATES INC.
proceedings
.com

Some format issues inherent in the e-media version may also appear in this print version.

Printed by Curran Associates, Inc. (2018)

Institute of Physics
Dirac House, Temple Back
Bristol BS1 6BE UK

Phone: 44 1 17 929 7481
Fax: 44 1 17 920 0979

techtracking@iop.org

Additional copies of this publication are available from:

Curran Associates, Inc.
57 Morehouse Lane
Red Hook, NY 12571 USA
Phone: 845-758-0400
Fax: 845-758-2633
Email: curran@proceedings.com
Web: www.proceedings.com

Volume 1081

2018

4th Euro-Mediterranean Conference on Materials and Renewable Energies (EMCMRE-4)
8–11 May 2017, Marrakech, Morocco

Accepted papers received: 2 August 2018
Published online: 10 September 2018

Preface

Papers

4th Euro-Mediterranean Conference on Materials and Renewable Energies (EMCMRE-4), May, 8-11, 2017, Marrakech-Morocco
IOP Publishing
IOP Conf. Series: Journal of Physics: Conf. Series **1081** (2018) 011001 doi:10.1088/1742-6596/1081/1/011001

Preface

The 4th Euro-Med Conference for Materials and Renewable Energies (EMCMRE-3) was held in Marrakech- Morocco Mai 8-11 2016. This conference is the 4th of a series that aims to strengthen relationships between material scientists from all sides of the Mediterranean Sea. The scientific topics of the EMCMRE conference series are important to countries surrounding the Mediterranean Sea with particular attention to novel problems using state of the art experimental techniques as well as advanced theoretical and computational approaches for materials science. This conference in the series is dedicated to the three main aspects of Renewable Energies, namely storing energy, spectroscopies for advanced materials and energy conversion, solar and biofuels, theory and modeling as well as novel materials.
The topics of the EMCMRE conference series encompasses the branches of physics, chemistry and computer science that directly impact Renewable Energies. The targeted audience includes Professors, Researchers and Engineers from Universities, Engineering Schools and Research Institutes. The participation of PhD and Masters Students from both sides of the Mediterranean Sea is strongly encouraged and supported with travel grants. The conference includes oral and poster presentations as well as plenary and keynote presentations given by high profile scientists, from all over the world, that will cover the key themes of the conference.

Editorial committee:

Prof. Hamid Oughaddou, France
Prof. Abdelkader Kara, USA
Dr. Azzedine Bendounan, France
Dr. Karima Lasri, USA
Dr. Andrew Mayne, France
Prof. Abedlilah Kaddouri, Morocco
Prof. Nabil Rochdi, Morocco
Prof. Abdelaziz El Boujlaidi, Morocco
Prof. Jean-Louis Lemaire, France
Dr. Khalil Amine, USA
Prof. Laurène Tétard, USA
Prof. Ismael Saadoune, Morocco

Keynotes and invited speakers for the EMCMRE-4 are:

Energy Storage
Keynote speaker: Prof. Yi Cui *(Stanford University, USA)*

Invited speakers:
Prof. Khalil Amine *(Argonne National Laboratory, USA)*
Prof. Guohua Chen *(University of Science and Technology, Hong Kong)*
Prof. Stanley Whittingham *(SUNY Binghamton, USA)*
Prof. Xiao Qing Yang *(Brookhaven National Laboratory, USA)*
Dr. Filippo Maglia *(BMW, Germany)*
Dr. Ali Abouimrane *(Qatar Environment and Energy Research Institute, Qatar)*
Dr. Jun Lu *(Argonne National Laboratory, USA)*
Prof. Jean Le Bideau *(University of Nantes, France)*
Prof. Abdelilah Benyoussef *(Mohammed V University, Morocco)*

IOP Conf. Series: Journal of Physics: Conf. Series **1081** (2018) 011001 doi:10.1088/1742-6596/1081/1/011001

Spectroscopies for Advanced Materials and Energy Conversion

Keynote speaker: Dr. Moulay Tahar Sougrati *(CNRS, France)*

Invited speakers:

Dr. Antonella Iadecola *(Soleil Synchrotron, France)*
Dr. Phoebe K. Allan *(University of Cambridge, United Kingdom)*
Prof. Lorenzo Stievano *(Montpellier University, France)*
Prof. Hervé Martinez *(Pau University, France)*
Prof. Michaël Deschamps *(Orléans University, France)*
Prof. Fouad Ghamouss *(François Rabelais University, France)*
Prof. Patrik Johansson *(Chalmers University of Technology, Sweden)*
Prof. Laurène Tetard *(University of Central Florida, USA)*

Solar and Biofuels

Keynote speaker: Prof. Abdou Lachgar *(Wake Forest University, USA)*

Invited speakers:

Prof. Scott M. Geyer *(Wake Forest University, USA)*

Theory and Modeling

Invited speakers:

Prof. Natalie A. W. Holzwarth *(Wake Forest University, USA)*
Prof. Moncef Said *(University of Monastir, Tunisia)*
Dr. Ari Paavo Seitsonen *(ENS-CNRS, France)*
Prof. Benkhettou Nour-Eddine *(Djillali Liabes University, Algeria)*
Dr. Andres Saul *(CINaM-CNRS, France)*

Novel Materials

Keynote speaker: Prof. Daniel Malterre *(University of Lorraine, France)*

Invited speakers:

Prof. Giovanni Marletta *(University of Catania., Italy)*
Prof. Thomas Greber *(Zurich University, Switzerland)*
Dr. Bernard Aufray *(CINaM-CNRS, France)*
Prof. Mustapha Ait Ali *(University Cadi Ayad, Morocco)*
Prof. Michael Altman *(Hong Kong University of Science and Technology)*
Dr. Geoffroy Prévot *(INP-CNRS, France)*

Peer review statement

All papers published in this volume of *Journal of Physics: Conference Series* have been peer reviewed through processes administered by the proceedings Editors. Reviews were conducted by expert referees to the professional and scientific standards expected of a proceedings journal published by IOP Publishing.

IOP Conf. Series: Journal of Physics: Conf. Series **1081** (2018) 012001 doi:10.1088/1742-6596/1081/1/012001

$Li_4Ti_5O_{12}$ powders by spray-drying: influence of the solution concentration and particle size on the electrochemical properties

Claire Jamin, Thomas Jungers, Caroline Piffet, Abdelfattah Mahmoud, Rudi Cloots, Bénédicte Vertruyen, Frédéric Boschini

GREENMAT, CESAM Research Unit, B6a, University of Liège, Sart-Tilman, 4000 Liège, Belgium

*Corresponding author: phone: +32 4 366 3532; e-mail: frederic.boschini@ulg.ac.be

Abstract. The lithium battery electrode compound $Li_4Ti_5O_{12}$ was synthesized by calcination of precursor powders obtained through spray-drying of solutions prepared with titanium isopropoxide and lithium nitrate. X-ray diffraction and thermal analysis coupled to mass spectrometry show that single phase crystalline $Li_4Ti_5O_{12}$ particles can be obtained after calcination at 800 °C for 2 hours. Decreasing the solution concentration leads to smaller particle sizes but also to an unexpected decrease of the electrochemical capacity, probably related to the presence of residual Li_2TiO_3. On the contrary, the capacity of the $Li_4Ti_5O_{12}$ powder prepared with the high concentration solution can be increased from 150 mAh/g to 165 mAh/g (C/4 rate) by grinding. These results highlight the fact that smaller particles do not systematically display better performances for Li^+ intercalation/desintercalation and confirm the need for a comprehensive approach including parameters such as crystallinity, phase purity or agglomeration.

1. Introduction

As a negative electrode material in Li-ion batteries, lithium titanate $Li_4Ti_5O_{12}$ has a theoretical capacity of 175 mAh g^{-1} and a voltage plateau at ~1.55 V vs. Li^+/Li^0. Its excellent lithium intercalation/desintercalation reversibility is favored by the fact that there is almost no volume change during charge–discharge cycling [1,2]. Major advantages of $Li_4Ti_5O_{12}$ are low toxicity and low cost. Besides, $Li_4Ti_5O_{12}$ is the electrode material of choice when maximum safety is required, because Li^+ ions are intercalated at less negative potentials than in the usual carbon-based anodes [3,4], preventing lithium plating at high cycling rates [5]; decomposition of the electrolyte and formation of a passivation layer (solid-electrolyte interface – SEI) at the surface of the electrode are also prevented [6].

Many synthesis methods [7-13] have been investigated in the literature as alternatives to the standard solid state technique, targeting lower synthesis temperature, shorter reaction time and/or smaller particle size [14-16]. The spray-drying method [17] is particularly attractive because it has a good reproducibility and is easily upscalable to industrial level. Spray-drying takes place by atomization of a suspension or solution into small droplets injected in a drying chamber with a hot air flow. After solvent vaporization, the resulting solid particles are separated from the air flow and collected in a container. When applied to suspensions of pre-existing insoluble solid powders, the spray drying technique is used to prepare dry granules of agglomerated powder. Several authors have used the spray drying of suspensions as an alternative to manually or mechanically mixing the reactants (typically a lithium salt and either anatase powder or in-situ generated TiO_2 colloidal particles) before solid state synthesis [18-23]. In the present work, we focus on spray drying of solutions, where excellent homogeneity is expected due to coprecipitation by solvent vaporization inside each individual droplet. This approach has been investigated by Yoshikiwa et al. [5] and Wu et al. [15,24,25] who studied the influence of the Li/Ti ratio and the temperature of heat treatment. Using the related technique of spray pyrolysis (where droplets are projected in a heated furnace), other authors [14] observed that modifications of the solution concentration affect the (micro)structural characteristics and the electrochemical performance of the $Li_4Ti_5O_{12}$ powders. It is therefore important to study how the solution concentration influences the synthesis of $Li_4Ti_5O_{12}$ by spray drying.

In the present work, $Li_4Ti_5O_{12}$ powders were prepared by spray-drying of solutions with concentrations ranging from 0.2 mol/l to 0.02 mol/l. The paper is divided in 3 parts, investigating respectively (1) the transformation of the as-sprayed powder into $Li_4Ti_5O_{12}$, (2) the influence of the solution concentration on the structure, microstructure and electrochemical capacity and (3) the influence of a further grinding step on the particle size and cycling performance.

2. Experimental

$Li_4Ti_5O_{12}$ powders were synthesized by heat treatment of precursor powders obtained by spray-drying of solutions with different concentrations: samples labelled LTO1, LTO2, LTO3 and LTO4 were obtained from solutions with titanium concentrations of 0.2, 0.1, 0.05 and 0.02 mol/l, respectively.

The solutions were prepared by dissolving 21.8 mmol titanium tetraisopropoxide (TTIP - Across, 99 % purity) and 17.4 mmol $LiNO_3$ (Aldrich, 99 % purity) in 6.5 ml isopropanol (VWR, HPLC 99 % purity) and 14 ml glacial acetic acid (Fisher, analytical grade); deionised water was subsequently added to the solutions under stirring to reach the final concentrations. Spray drying was performed in a lab-scale apparatus Büchi B-191 through a co-current atomizer (i.e., sprayed solution and drying air flow in the same direction) using a 0.7 mm nozzle. The inlet temperature was 120 °C with a liquid feed rate of 1.4 ml/min. The process was carried out in air, with atomizing pressure of 1 bar and a hot air flow rate of 700 normal l/h. The resulting powders were then calcined for 2 h in air at various temperatures from 200 °C to 800 °C.

Thermal analysis (thermogravimetry – TGA and differential scanning calorimetry – DSC) was performed with a Setaram TG-DSC111 system coupled and synchronized to a mass spectrometer (Pfeiffer Omnistar) for analysis of the evolved gases; powders were placed in alumina crucibles and heated in flowing air with a heating rate of 5 °C/min. X-ray diffraction patterns were recorded with a Siemens D5000 powder diffractometer (CuKα radiation, 15-65°2θ, 0.04° step size, 2.2 s/step). The TOPAS software was used for Rietveld refinement, using the fundamental parameters approach to model the instrumental contribution [26]. Fourier transform infrared (FTIR) spectroscopy was performed on powders dispersed in KBr disks. The powder morphology was characterized by scanning electron microscopy (Philips ESEM XL30 FEG). The particle size distribution was measured on isopropanol suspensions of calcined powders with a Mastersizer 2000 – Hydro 2000S apparatus (Malvern). Ball milling was performed at 200 rpm with a planetary crusher (Retsch PM 400/2), using a 7.5 wt% suspension in isopropanol. Electrochemical measurements were carried out with Swagelok two-electrode coin cells assembled in an argon glove box. 13 mm-diameter pellets were prepared by pressing, in a stainless steel grid, ca. 4 mg of a mixture of $Li_4Ti_5O_{12}$, acetylene black and polyvinylidene fluoride (PVDF) binder in 80:10:10 weight ratio. The counter electrode was metallic lithium, the current collectors were stainless steel and the separator between electrodes was a 25 μm monolayer polypropylene membrane (Celgard). The electrolyte was 1 mol/l $LiPF_6$ in ethylene carbonate:dimethylcarbonate (1/1 v/v). Galvanostatic discharge-charge curves were measured using a PAR 263A potentiostat, from 2 V to 1 V vs. Li^+/Li^0.

IOP Conf. Series: Journal of Physics: Conf. Series **1081** (2018) 012001 doi:10.1088/1742-6596/1081/1/012001

3. Results and discussion

3.1. From as-sprayed powder to $Li_4Ti_5O_{12}$

The powders obtained by spray-drying are amorphous and have to be calcined to induce crystallization. The formation of $Li_4Ti_5O_{12}$ from the LTO1 as-sprayed powder was studied by a combination of thermal analysis, FTIR spectroscopy and X-ray diffraction.

Thermogravimetry (TG) and differential scanning calorimetry (DSC) data were collected with a TG–DSC system coupled to a mass spectrometer (MS) for identification of the evolved gases. Figure 1 shows the TG curve, the DSC curve and the MS signals corresponding to m/z ratios of 18 (H_2O), 30 (NO) and 44 (CO_2). The thermal decomposition of the as-sprayed powder can be divided in three temperature ranges according to the inflexion points in the TG curve. Below 200 °C, the mass loss is due to endothermic release of water, corresponding to vaporization of loosely adsorbed water and decomposition of the –OH groups from the hydrolyzed titanium species. Between 200 °C and 400 °C, the exothermic progressive oxidation of organics into CO_2 and H_2O is accompanied by a first decomposition of nitrates. Finally, residual nitrates are decomposed between 450 °C and 550 °C, as confirmed by FTIR spectra (not shown) where the characteristic nitrate peak at 1380 cm^{-1} is still present in the sample heated to 400 °C and disappears after heating to 500 °C.

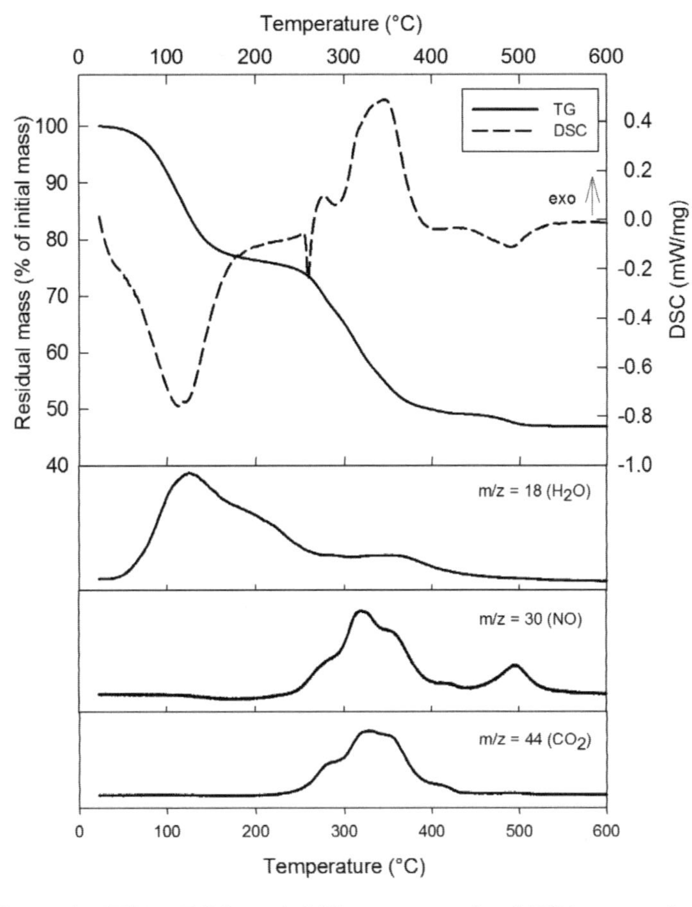

Figure 1. TGA, DSC and MS response for LTO1 spray-dryed powder, measured under air (5°C min^{-1}). The MS response related to H_2O, NO and CO_2 (m/z = 18, 30 and 44, respectively) is presented.

Figure 2 presents the XRD patterns of the LTO1 powders heated for 2 hours at temperatures from 200 °C to 800 °C. The onset of crystallization is observed after 2 hours at 300 °C with the presence of a weak diffraction peak corresponding to the (101) reflection of the anatase polymorph of TiO_2. After 2 hours at 500 °C, the reflections of the Li_2TiO_3 phase are clearly present, together with reflections from both the anatase and rutile polymorphs of TiO_2. The anatase-to-rutile transition is completed after 2 hours at 600 °C. After 2 hours at 700 °C, $Li_4Ti_5O_{12}$ becomes the major phase and increasing the temperature to 800 °C yields $Li_4Ti_5O_{12}$ without apparent secondary phase. These results show that $Li_4Ti_5O_{12}$ is formed by reaction between TiO_2 and Li_2TiO_3. Crystallization of Li_2TiO_3 and TiO_2 intermediate phases is commonly observed when $Li_4Ti_5O_{12}$ is synthesized with titanium alkoxide reactants because TiO_2 nuclei are easily formed by condensation of the hydrolyzed titanium species [5].

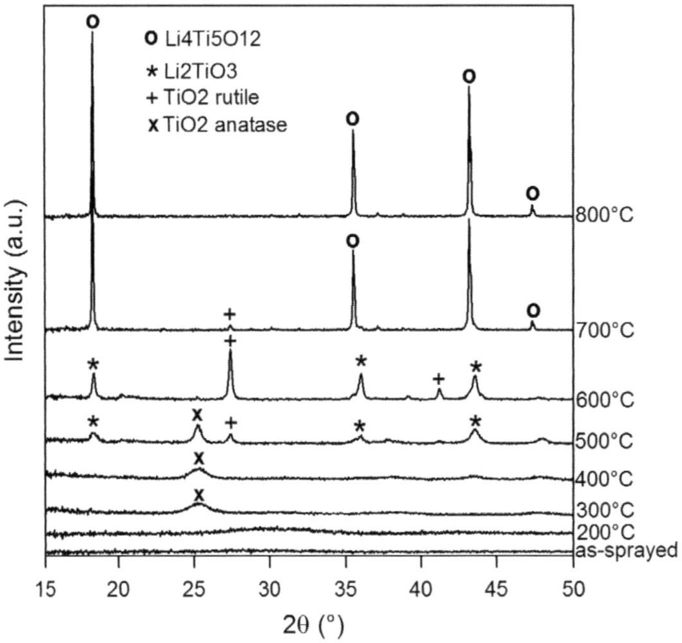

Figure 2. XRD patterns of LTO1 heated for 2 hours at temperatures from 200 °C to 800 °C.

3.2. Influence of the solution concentration

Figure 3 shows how a decrease of the solution concentration from 0.2 mol/l (LTO1) to 0.1 mol/l (LTO2), 0.05 mol/l (LTO3) or 0.02 mol/l (LTO4) affects the crystallinity and microstructure of the powders heated at 800 °C for 2 hours. The XRD patterns in the left column of figure 3 indicate that the $Li_4Ti_5O_{12}$ main phase is contaminated by a residual Li_2TiO_3 phase in the case of the powders prepared using the two most dilute solutions. The Li_2TiO_3 contents estimated by Rietveld refinement are 3 wt% and 15 wt% in the LTO3 and LTO4 powders, respectively. The absence of TiO_2 reflections (not shown) suggests that the $Li_4Ti_5O_{12}$ and Li_2TiO_3 phases might be slightly off-stoichiometric ($Li_{4-x}Ti_5O_{12-x/2}$ and $Li_{2-y}TiO_{3-y/2}$), due to incomplete diffusion of lithium from Li_2TiO_3 into TiO_2.

Scanning electron micrographs displaying the microstructures of the calcined powders are shown in the middle column of figure 3. In all cases, the solidified crusts of the droplets have shapes very different from spherical hollow spheres, indicating that buckling of the droplets occurred during spray drying; such a behavior is frequently reported in the literature [17,27] and there have been many attempts to model the mechanical instabilities in drying droplets [17,28]. The SEM micrographs also suggest some agglomeration of the dried droplets. In order to probe the strength of these agglomerates, the particle size distributions were measured before and after ultrasonication and are shown in the right column of figure 3. The agglomerates are characterized by $d_{0.5}$ values of about 20 µm in all powders. Ultrasonication shifts the $d_{0.5}$ down to values between 3.5 and 5.5 µm except for the LTO4 powder with a $d_{0.5}$ that does not go below 8 µm. Comparing these values with the apparent sizes in the SEM micrographs, these results suggest the formation of stronger agglomerates in the case of powders synthesized with lower concentrations of precursor solutions.

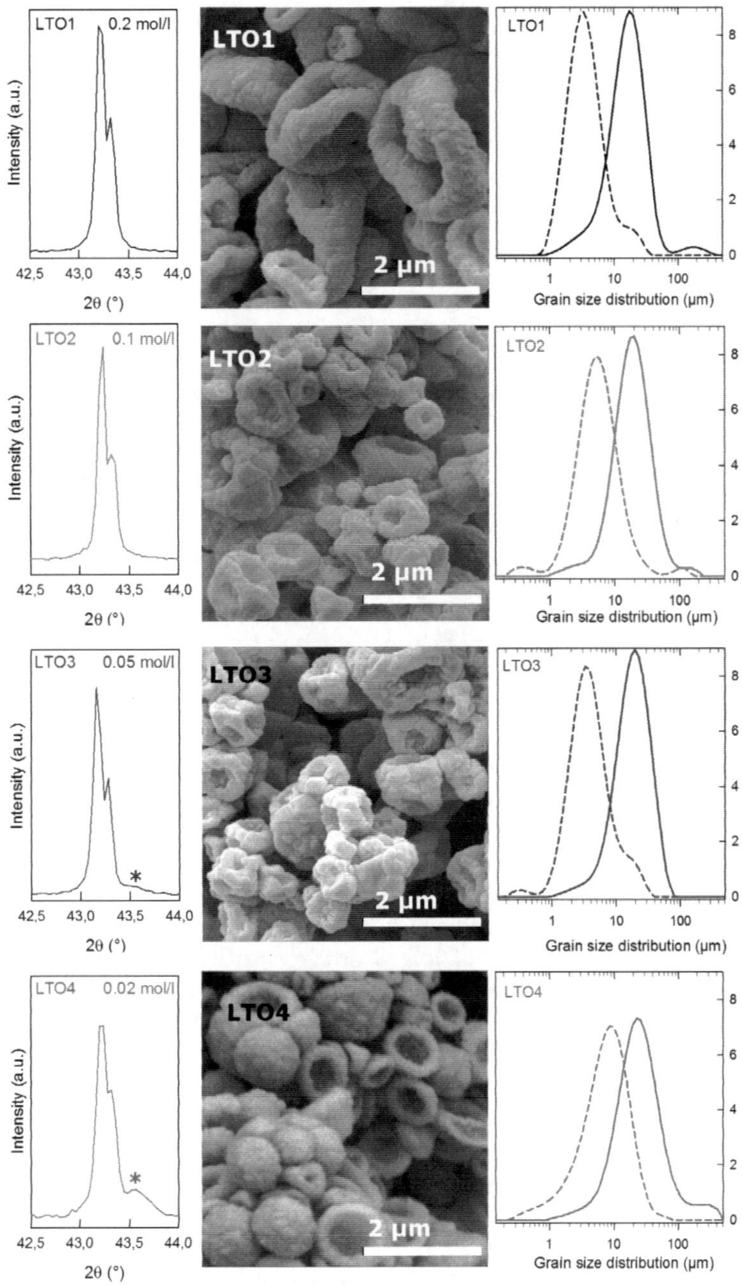

Figure 3. Comparison of powders heated at 800°C for 2 hours: (left) 42.5–44.0° 2θ range in the XRD pattern with Li_2TiO_3 secondary phase highlighted by an asterisk; (middle) scanning electron micrographs; (right) particle size distributions.

The electrochemical properties of the powders heated at 800 °C for 2 h were characterized by galvanostatic cycling at C/4 rate. Figure 4 presents the second discharge curves with capacity values quoted per gram of powder. The best capacities (~ 150 mAh g^{-1}) are obtained for the LTO1 and LTO2 powders. The LTO3 and LTO4 powders have lower capacities, 110 and 103 mAh g^{-1} respectively. These two powders contain residual Li_2TiO_3 phase, which is electrochemically inert [29,30]. However, the decrease in capacity is by ~30 % while the Li_2TiO_3 phase represents max. 15 wt%. This suggests that the grain size distribution and especially the formation of agglomerates contributes to the reduction of the electrochemical capacity, so that decreasing the solution concentration in the hope of obtaining smaller particles does not lead to the expected improvement of the electrochemical performance. In summary, the best results were obtained for the powders prepared from the solutions with the highest concentrations. The last section of this paper investigates whether a further improvement can be obtained by grinding the powder.

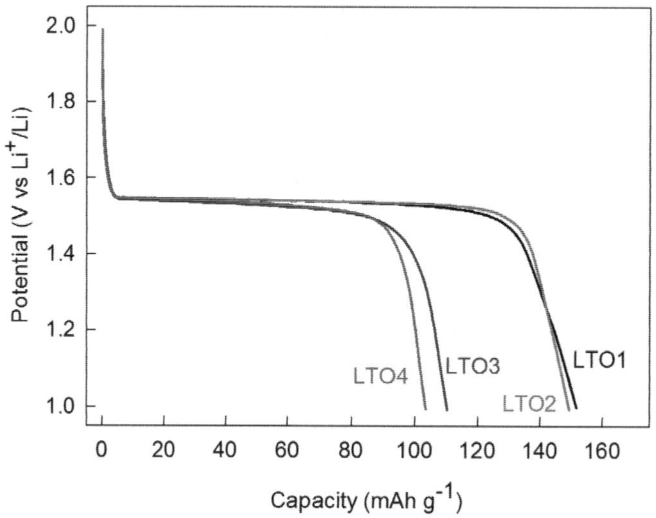

Figure 4. Discharge curves at C/4 rate in the voltage range 1.0–2.0 V.

3.3. *Influence of grinding*

To investigate the effect of grain size on the electrochemical properties, the LTO1 powder was ground for up to 4 hours. Figure 5 shows the SEM micrographs of the powder before grinding and after 30 min, 1 h or 4 h grinding. After 30 minutes or 1 hour of grinding, the majority of the dried droplets were broken into smaller pieces made up of joined primary particles. 4 hours of grinding leads to further reduction of the size of the primary particles into nanoparticles. The bottom of figure 5 presents the second discharge curves at C/4 rate for the LTO1 powder before and after grinding. The capacity tends to increase with decreasing particle size, from 152 mAh g^{-1} (before grinding) to 165 mAh g^{-1} (after 4 h of grinding). The shape of the curve before and after the plateau also changes with grinding, especially for the powder ground during 4 hours. This may be related to the partial amorphisation detected as a decrease of the XRD signal intensity, with the area under the main $Li_4Ti_5O_{12}$ reflection decreasing by 15 % after 1 h and 45 % after 4 h (not shown). Borghols and al. have suggested that this voltage profile, typical for nanoparticles, might be the consequence of strain and/or interface energy, which has relatively more impact in smaller particles [31].

Figure 6 compares the electrochemical capacity retention of the LTO1 unground powder with the LTO1 powder ground for 4 hours, assessed by measuring 30 galvanostatic charge/discharge cycles at C/4 rate and 10 galvanostatic charge/discharge cycles at C, 4C, 10C and C/4 rates. The behavior of both powders is qualitatively similar, with a good capacity retention at C/4 rate after the first few cycles and the usual decrease in capacity at faster cycling speeds (C, 4C, 10C). Quantitatively, the capacities are systematically higher in the case of the ground powder, due to the larger surface area resulting from the smaller grain size.

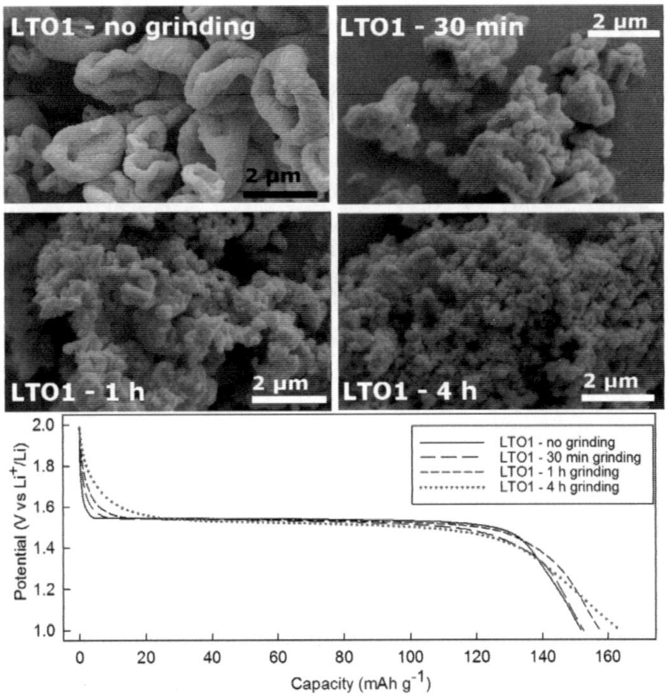

Figure 5. SEM micrographs and discharge curves at C/4 rate of the LTO1 powder before grinding and after 30 min, 1 h or 4 h grinding.

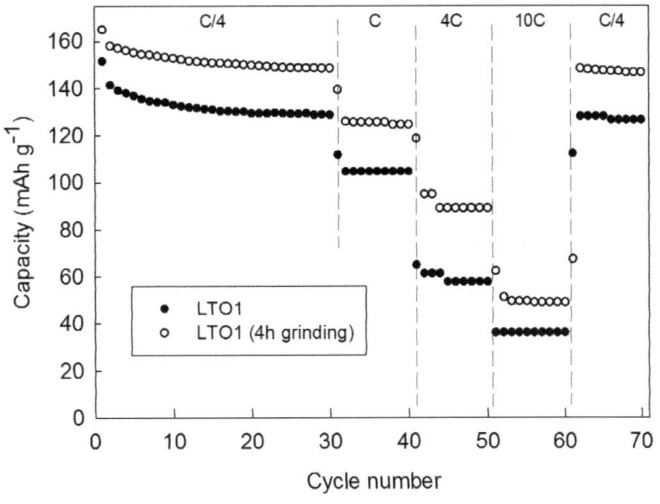

Figure 6. Rate capability of LTO1 without grinding and after 4 h grinding.

4. Conclusion

$Li_4Ti_5O_{12}$ was synthesized by spray drying of solutions with several concentrations. The $Li_4Ti_5O_{12}$ phase is formed at 700 °C and above by reaction between TiO_2 and Li_2TiO_3 intermediary phases in the precursor powders. The morphology and agglomeration of the particles after 2 h at 800 °C varies according to concentration of the precursor solution. Decreasing the solution concentration leads to smaller particles but residual Li_2TiO_3 is present and the electrochemical capacity tends to decrease. The best electrochemical performance is displayed by the powder prepared from the high concentration solution and can be further improved by grinding to reach 165 mAh/g at C/4 rate.

5. References

[1] Yi T F, Jiang L J, Shu J, Yue C-B, Zhu R S and Qiao H B 2010 *J. Phys. Chem. Solids* **71** 1236–42

[2] Ohzuku T, Ueda A and Kouguchi M 1995 *Electrochem. Soc.* **142** 1431–5

[3] Hayner C M, Zhao X and Kung H H 2012 *Annu. Rev. Chem. Biomol. Eng.* **3** 445–71

[4] Gao J, Jiang C and Wan C 2010 *Ionics* **16** 417–24

[5] Yoshikawa D, Kadoma Y, Kim J M, Ui K, Kumagai N, Kitamura N and Idemoto Y 2010 *Electrochem. Acta* **55** 1872–9

[6] Zaghib K, Dontigny M, Guerfi A, Charest P, Rodrigues I, Mauger A and Julien C M 2011 *J. Power Sources* **196** 3949–54

[7] Prakash A S, Manikandan P, Ramesha K, Sathiya M, Tarascon J M and Shukla A K 2010 *Chem. Mater.* **22** 2857–63

[8] Jung H G, Kim J, Scrosatia B and Suna Y-K 2011 *J. Power Sources* **196** 7763–6

[9] Li J, Jin Y L, Zhang X G and Yang H 2007 *Solid State Ionics* **178** 1590–4

[10] Cheng L, Yan J, Zhu G N, Luo J-Y, Wang C X and Xia Y Y 2010 *J. Mater. Chem.* **20** 595–602

[11] Ogihara T, Yamada M, Fujita A, Akao S and Myoujin K 2011 *Mater. Res. Bull.* **46** 796–800

[12] Zou H, Zhang G and Shen P K 2010 *Mater. Res. Bull.* **45** 149–52

[13] Jamin C, Traina K, Eskenazi D, Krins N, Cloots R, Vertruyen B and Boschini F 2013 *Mater. Res. Bull.* **48** 4641–6

[14] Ju S H and Kang Y C 2009 *J. Power Sources* **189** 185-90

[15] He Z, Wang Z, Wu F, Guo H, Li X and Xiong X 2012 *J. Alloys Compd.* **540** 39–45

[16] Zhang Q, Peng W, Wang Z, Li X, Xiong X, Guo H, Wang Z and Wu F 2013 *Solid State Ionics* **236** 4016–22

[17] Nandiyanto A B D and Okuyama K 2011 *Adv. Powder Technol.* **22** 1–19

[18] Zhu G N, Liu H J, Zhuang J H, Wang C X, Wang Y G and Xia Y Y 2011 *Energy Environ. Sci.* **4** 4016–22

[19] Han S W, Ryu J H, Jeong J and Yoon D H 2013 *J. Alloys Compd.* **570** 144–9

[20] Hsieh C T, Chen I L, Jiang Y R and Lin J Y 2011 *Solid State Ionics* **201** 60–7

[21] Hsiao K C, Liao S C and Chen J M 2008 *Electrochem. Acta* **53** 7242–7

[22] Lee B and Yoon J R 2013 *Electron. Mater. Lett.* **9** 871–3

[23] Han S W, Shin J W and Yoon D H 2012 *Ceram. Int.* **38** 6963–8

[24] Wu F, Wang Z, Li X, Guo H, Yue P, Xiong X, He Z and Zhang Q 2012 *Electrochim. Acta* **78** 331–9

[25] Wu F, Li X, Wang Z, Guo H, He Z, Zhang Q, Xiong X and Yue P 2012 *J. Power Sources* **202** 374–9

[26] Cheary R W and Coelho A 1992 *J. Appl. Crystallogr.* **25** 109–21

[27] Ståhl K, Claesson M, Killiehorn P, Lindén H and Bäcksöm K 2002 *Int. J. Pharm.* **233** 227–37

[28] Tsapis N, Dufresne E R, Sinha S S, Riera C S, Hutchinson J W, Mahadevan L and Weitz D A 2005 *Phys. Rev. Lett.* **94** 018302

[29] Soares A, Fraisse B, Morato F, Ionica-Bousquet C M and Monconduit L 2012 *J. Power Sources* **208** 440–6

[30] Lu J, Peng Q, Wang W, Nan C, Li L and Li Y 2013 *J. Am. Chem. Soc.* **135** 1649–52

[31] Borghols W J H, Wagemaker M, Lafont U, Kelder E M and Mulder F M 2009 *J. Am. Chem. Soc.* **131** 17786–92

Acknowledgements
The authors thank A. Zubiaur for the thermal analysis measurement. CJ thanks FRIA (Belgium) for a PhD grant. TJ thanks FRS-FNRS (Belgium) for a PhD grant. AM thanks the Walloon region for a Beware Fellowship Academia (2015-1, RESIBAT n° 1510399). The authors are grateful to Walloon Region for financial support through the BATWAL project (PE Plan Marshall 2.vert).

Spin transport and precession in semiconductors in the drift-diffusion regime

Nabil Rochdi[1,3] **and Viatcheslav Safarov**[2]
[1] SIAM, Faculty of Sciences Semlalia, Cadi Ayyad University, Prince Moulay Abdellah avenue, PO Box 2390, 40000 Marrakesh - Morocco
[2] PMC, Ecole Polytechnique, route de Saclay, 91128 Palaiseau Cedex - France

E-mail: nabil.rochdi@uca.ma

Abstract. We present a theoretical analysis of spin transport in semiconductors in the drift-diffusive regime. We investigated spatial distributions of spin polarization and spin current under electrical injection and in presence of magnetic field. Numerical evaluations made for silicon highlight its strong potential for spintronic applications: the electron spin length at room temperature can spread over several microns and a rather weak magnetic field (of the order of 10 mT) can strongly modify the injected spin distribution in the structure.

1. Introduction

Spintronics have received considerable attention last decades thanks to its innovative potential applications in many fields such opto- and nanoelectronic engineering, quantum computation and data storage. It promises a reduction in energy consumption, high-speed operation and easy device integration. With recent advances in various nanofabrication techniques [1,4], much work has been devoted to spintronics taking aim at the elaboration of reliable devices such as Magnetic Random Access Memories (MRAMs).

The approach of spintronics involves the intrinsic angular momentum of conduction electrons (spins) as an additional degree of freedom to generate new functionalities of devices [5-10]. In those lines, Datta and Das proposed a device in which spin polarized electrons are electrically injected from a ferromagnetic material into a semiconductor, drift along the device without losing their spin orientation, and are finally collected by the same kind of ferromagnetic material/semiconductor junction [11]. One can manipulate the electrons spin for information processing by applying some perturbation such as an external magnetic field or another equivalent effect (Rashba effect for example). By virtue of its operating principle, such device is commonly called "spin transistor". Thus, the key challenges for the spin transistor achievement lie in the following points: spin injection from ferromagnetic metals to semiconductors, spin transport in semiconductors, spin processing, and spin detection.

[3] To whom any correspondence should be addressed.

Spin injection from a ferromagnetic material into a semiconductor is not a trivial issue. The Schottky barrier formed by the metal-semiconductor contact prevents the injection of electrons from the metal into the conduction band of the semiconductor. To overcome this issue, the insertion of a thin insulating layer between the ferromagnetic metal and the semiconductor was putted forward in order to unpin the electrochemical potential at the ferromagnetic metal/semiconductor interface [12,13]. In these structures, the spin-polarized electrons get through this insulating layer by tunneling. Moreover, the insulating layer could act as a physical barrier against the metal diffusion in the semiconductor, preventing then the formation of an interfacial dead layer able to lead up to spin depolarization. The introduction of a tunnel barrier helps to resolve the mismatch between the conductivities of the two materials, which strongly affects the current polarization in ferromagnetic metal/normal metal junctions [14]. The use of tunneling barrier paved the way for successful electrical spin injection into semiconductors [15-17], stimulating then many studies on the elaboration of nanoscaled thin oxide layers [18-21].

Spin transport and processing rely heavily on the spin length, namely the distance along which polarized electrons can travel without losing their spin orientation. The spin length depends on the semiconductor material, or more precisely on the spin relaxation time in it. As a matter of fact, the spin length limits the size of spintronic devices; it must exceed the distance between the spin injector and the spin detector in case of the spin transistor.

In the present paper, we focus on theoretical analysis of room temperature (RT) spin transport and spin processing by external magnetic field in semiconductors in pure drift-diffusion regime. The numerical evaluations made in the framework of the developed model for silicon highlights a high potential in spintronics of this fundamental material of modern microelectronics.

2. Analysis of spin transport and precession

We assume that, through the ferromagnetic/insulator/semiconductor junction, we inject in semiconductor unequal electrical current densities J_+ and J_- of electrons with spin-up and spin-down orientations respectively, and whose sum corresponds to the total current density J. These densities give rise to the current polarization P_J as given by the following equation:

$$P_J = \frac{J_+ - J_-}{J_+ + J_-} = \frac{\Delta J}{J} \tag{1}$$

It's essential to understand what is meant by the injected current densities J_+ and J_-: Despite the fact that they refer to electrons current densities, these quantities are positive and correspond to modulus of spin-up and spin-down electrons current densities vectors ($\vec{J}_+$ and $\vec{J}_-$) that are oriented from the injector toward the detector.

Further, the carrier spin polarization P is characterized by the respective concentrations, n_+ and n_-, of electrons with spin-up and spin-down orientations:

$$P = \frac{n_+ - n_-}{n_+ + n_-} = \frac{\Delta n}{n} \tag{2}$$

Where n is the total carrier concentration.

In this theoretical study, we consider that injected current densities $\vec{J}_+$ and $\vec{J}_-$ as well as the electrical field into the semiconductor $\vec{E}$ are oriented toward the y-axis as shown in figure 1.

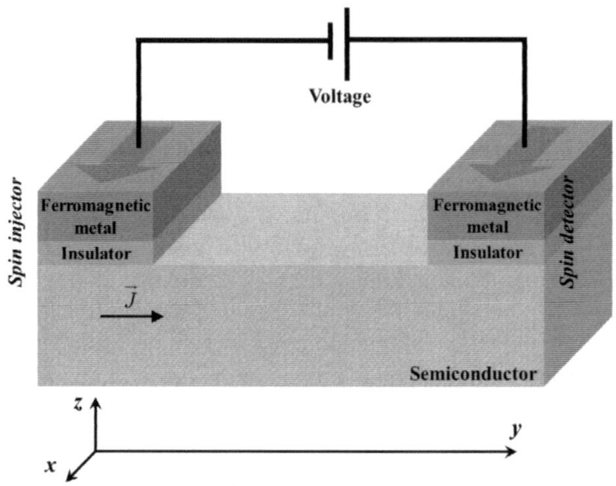

Figure 1. Schematic of a semiconductor-based spin transistor structure consisting of spin injector and detector with ferromagnetic metal/insulator/semiconductor junctions. The currents of injected spin-polarized electrons and the electrical field in the semiconductor are oriented along the y-direction.

2.1. Drift-diffusion spin transport

Assuming that the semiconductor material is free of magnetic impurities, the mobility μ and the diffusion coefficient D of carriers are spin-independent and the difference in conductivities of electron gases with spin-up and spin-down orientations is given only by the difference in their concentrations. Thus, for n-type semiconductors, the spin transport is driven by the standard drift-diffusion and continuity equations for electrons with spin-up and spin-down orientations [22,23] as following:

$$\vec{J}_\pm = \sigma_\pm \vec{E} + qD \cdot \vec{\nabla} n_\pm \tag{3}$$

$$\frac{\partial n_\pm}{\partial t} = -\left(\frac{1}{q} \vec{\nabla} \vec{J}_\pm \quad \frac{n_+ \pm n}{2\tau_S} \right) \tag{4}$$

Where σ_+ and σ_- stand for respective conductivities of spin-up and spin-down polarized electrons, q represents the elementary charge ($q = +1.6\ 10^{-19}$ C) and τ_S is the spin relaxation time.

In the case of weak injection, n could be regarded as constant everywhere in the semiconductor ($\vec{\nabla} n = \vec{0}$), and the diffusion contribution to the total current is insignificant. This approximation is consistent if we assume that the semiconductor doping is homogenous and no local excitation leading to carriers' gradient is applied. Thus, in the stationary state ($\frac{\partial n}{\partial t} = \frac{\partial (\Delta n)}{\partial t} = 0$), the previous equations lead to the following one-dimensional equations:

$$D \frac{\partial^2 (\Delta n)}{\partial y^2} - \frac{J}{qn} \frac{\partial (\Delta n)}{\partial y} - \frac{\Delta n}{\tau_S} \quad 0 \tag{5}$$

$$\Delta J = \frac{\Delta n}{n} J \quad qD \frac{\partial(\Delta n)}{\partial y} \tag{6}$$

Equation (5) is equivalent to the drift-diffusion equation obtained in [22,23]. The latter usually involves the electric field inside the semiconductor, which is a quantity difficult to measure. Therefore, we prefer to characterize the electron motion in semiconductor by the total current density J, since it is easily measurable in experiment. Solutions of equations (5) and (6) are explicit and provide the stationary distributions of differences between spin-up and spin-down populations' concentrations Δn and current densities ΔJ as following:

$$\Delta n = A_1 \exp\left(-\frac{y}{l_S}\right) + A_2 \exp\left(\frac{y}{l_0}\right) \tag{7}$$

$$\Delta J = C_1 \exp\left(-\frac{y}{l_S}\right) + C_2 \exp\left(\frac{y}{l_0}\right) \tag{8}$$

Where A_1, A_2, C_1 and C_2 are constants depending on boundary conditions, and l_S and l_0 are characteristic spin lengths given by the expression:

$$l_{S,0} = \sqrt{D\tau_S}\left[\sqrt{\gamma^2+1} \pm \gamma\right] \tag{9}$$

The dimensionless quantity γ is defined by the expression:

$$\gamma = \frac{J}{2qn}\left(\frac{\tau_S}{D}\right)^{1/2} \tag{10}$$

From a physical point of view, the parameter γ is related to the ratio of electron drift velocity $v_{drift} = \frac{J}{qn}$ and electron spin diffusion velocity $v_{diff} = \left(\frac{D}{\tau_S}\right)^{1/2}$, or the ratio of electron drift length $l_{drift} = v_{drift} \cdot \tau_S$ and the spin diffusion length $l_{diff} = v_{diff} \cdot \tau_S$ as following:

$$\gamma = \frac{v_{drift}}{2 \cdot v_{diff}} = \frac{l_{drift}}{2 \cdot l_{diff}} \tag{11}$$

The lengths l_{drift} and l_{diff} correspond respectively to the distances over which the spin can drift and diffuse during the relaxation time τ_S.

In the case of a semi-infinite semiconductor extended along positive y-axis ($A_2 = C_2 = 0$), carrier spin and current polarizations decrease exponentially with a decay rate given by the characteristic spin length l_S. Thus, the found expression yields the electrons spin polarization P:

$$P = \frac{(\Delta n)_0}{n} \exp\left(-\frac{y}{l_S}\right) \tag{12}$$

Where $(\Delta n)_0$ corresponds to the difference between spin-up and spin-down concentrations at injection ($(\Delta n)_0 = \Delta n(y=0)$).

In the case of silicon, the RT spin length l_S is presented in figure 2 as a function of total current density of injected electrons J. The used value for τ_S is 10^{-8} s and it corresponds to that measured, at

IOP Conf. Series: Journal of Physics: Conf. Series **1081** (2018) 012002 doi:10.1088/1742-6596/1081/1/012002

ambient, for low-doped n-type silicon (in the range of 7.10^{14} cm^{-3} to 8.10^{16} cm^{-3}) by electron spin resonance experiments [24]. According to this figure, one can notice that, for weak injected current ($\gamma \ll 1$), the l_S value tends to the classical spin diffusion length l_{diff} which is about 5.5 μm (for $J = 0$ in figure 2). This result is consistent with experimental measurements showing that the spin can diffuse in silicon over several microns at ambient [7,8], and represents an excellent prospect since the typical feature sizes of silicon integrated circuits are ranging from several microns down to hundreds of nanometers. In addition, the spin length increases with stronger injected current ($\gamma \gg 1$) and tends to the asymptotic value of the spin drift length $l_{drift} = (\tau_s / qn) \cdot J$. This situation is equivalent to that described in [23] for strong electrical field into the semiconductor.

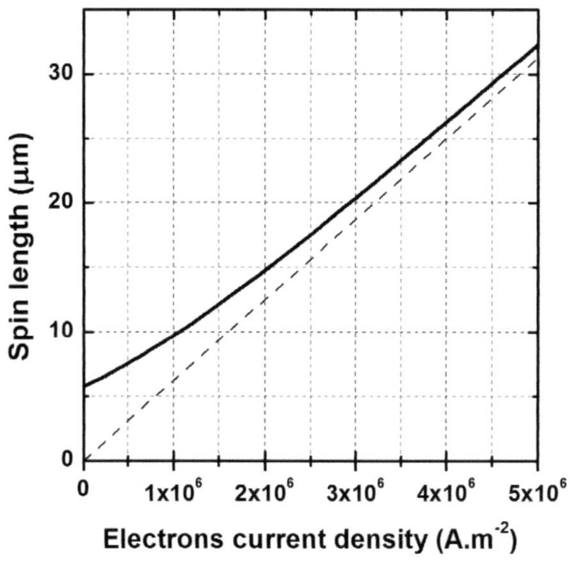

Figure 2. RT spin length in semi-infinite low-doped n-type silicon ($n = 10^{16}$ cm^{-3}) as a function of injected current. The dashed line corresponds to the spin drift length l_{drift}. Spin relaxation time is 10^{-8} s.

It should be emphasized that the electron spin polarization P and electron current polarization P_J do not have the same behavior. Taking into account the fact that the difference of conductivities between spin-up and spin-down electrons depends on the concentrations difference, the relationship between these polarizations is obtained according to equation (6):

$$P_J = P - \frac{D}{v_{drift}} \frac{\partial(P)}{\partial y} \tag{13}$$

Figure 3 shows distributions inside the semiconductor of spin and current polarizations, in the case of low-doped n-type silicon, for different injected current densities. The curves of figure 3 are obtained when normalizing the current density to that of injection: $P_J(y = 0) = 1$.

The left graph of figure 3, corresponding to a donor concentration of 10^{16} cm^{-3}, shows that spin and current polarizations increase with injected current since the latter increases the distance over which electrons can travel with weak spin attenuation. Indeed, apart from the concentrations difference between spin-up and spin-down electrons, the observed difference between P and P_J depends also on the difference of velocities of the two polarized electron populations. The figure 3, shows the distributions of electron polarizations under the weak injection hypothesis (n remaining constant over semiconductor) and for low donor doping concentrations (in the range corresponding to $\tau_S = 10^{-8}$ s as considered above). As shown in this graph, concentrations of spin-up (n_+) and spin-down (n_-) electrons are different at injection from the ferromagnetic metal and tend to even out because of spin relaxation ($P(y \to \infty) = 0$). From data of the figure 3, we can see the influence of the diffusion on differences between spin-up and spin-down currents and concentrations: After injection of spin-polarized electrons, the diffusion tends to enhance the current of majority-injected spins regarding the minority one, in contrast to the difference in concentrations, which is reduced.

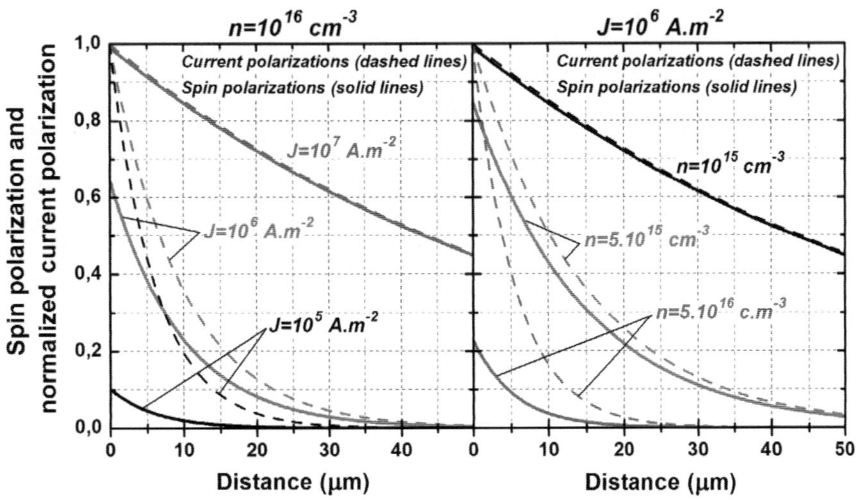

Figure 3. RT distributions of spin polarizations (in solid lines) and normalized current polarizations (in dashed lines) in n-doped silicon for different injected current densities (10^5 A.m^{-2}, 10^6 A.m^{-2}, and 10^7 A.m^{-2}) at n = 10^{16} cm^{-3} (left graph) and for different concentrations (10^{15} cm^{-3}, 10^{16} cm^{-3}, and 5×10^{16} cm^{-3}) at J = 10^6 A.m^{-2} (right graph). Spin relaxation time is 10^{-8} s.

2.2. Magnetic field-induced spin precession

We investigated injected electrons spin behavior and distribution when an external magnetic field is applied perpendicularly to the spin orientation. For this purpose, we introduce the mean spin vector of injected electrons defined as: $\vec{S} = \sum \vec{s}_i / n$, where $\vec{s}_i$ are spins of individual electrons.

Without magnetic field, the mean spin is given by the spin polarization as following:

$$S = \|\vec{S}\| = \frac{1}{n}\left[n_+\left(\frac{1}{2}\right) + n_-\left(-\frac{1}{2}\right)\right] = \frac{P}{2} \qquad (14)$$

In the presence of a magnetic field $\vec{B}$ non collinear to the injected spin orientation ($\vec{B} \times \vec{S} \neq \vec{0}$), the coupling between $\vec{B}$ and $\vec{S}$ leads to the precession of injected spins around $\vec{B}$ with the Larmor frequency $\vec{\Omega} = \frac{g\mu_B}{\hbar}\vec{B}$, where g, μ_B and $\hbar$ stand respectively for the g factor of conduction electrons, the Bohr magneton, and the reduced Planck constant. Moreover, applying an external magnetic field induces Lorentz forces that can modify the orbital motion of electrons. In the case of weak magnetic fields, we can assume that the contribution of Lorentz forces has a weak influence to be taken into account. Thus, inserting the coupling term in equation (5) leads to the following equation:

$$D\frac{\partial^2(\vec{S})}{\partial y^2} - \frac{J}{qn}\frac{\partial(\vec{S})}{\partial y} - \frac{1}{\tau_S}\vec{S} + \frac{g\mu_B}{\hbar}\vec{B}\times\vec{S} \quad 0 \tag{15}$$

Different orientations of magnetic field and injected spins are possible. However, in the case of the planar structure presented in figure 1, a rather weak in-plane magnetic field can perturb the magnetization of the ferromagnetic layer, for example by changing the orientation of magnetization from x- to y-orientation in the case of a magnetic field along the y-direction. To put magnetization perpendicular to the layer is much more difficult, a magnetic field stronger than 1 Tesla is needed. Therefore, the orientation of $\vec{B}$ parallel to z-axis is preferable: it can induce a rapid Larmor precession of spins inside the semiconductor without noticeable variation of magnetization in the ferromagnetic layer.

By resolving equation (15) in the case of a semi-infinite semiconductor [25] under an external magnetic field along the z-direction, we find the general expressions of the spin components S_X and S_Y as following:

$$\begin{cases} S_X = \frac{(\Delta n)_0}{2n}\exp\left(-\frac{y}{l_A}\right)\cos\left(\frac{y}{l_R}\right) \\ S_Y = \frac{(\Delta n)_0}{2n}\exp\left(-\frac{y}{l_A}\right)\sin\left(\frac{y}{l_R}\right) \end{cases} \tag{16}$$

Where the spin characteristic attenuation length l_A and rotation length l_R are described by the following expressions:

$$l_A = \frac{l_{diff}}{\left[\left(\gamma^2+1\right)^2 + \left(\tau_s\Omega\right)^2\right]^{1/4}\cos\left[\frac{1}{2}\arctan\left(\frac{\tau_s\Omega}{\gamma^2+1}\right)\right] - \gamma} \tag{17}$$

$$l_R = \frac{l_{diff}}{\left[\left(\gamma^2+1\right)^2 + \left(\tau_s\Omega\right)^2\right]^{1/4}\sin\left[\frac{1}{2}\arctan\left(\frac{\tau_s\Omega}{\gamma^2+1}\right)\right]} \tag{18}$$

Figure 4 illustrates schematically the precession of spin vector along the semiconductor channel in the case of an external magnetic field oriented along the z-axis. The determined solutions reduce, for zero magnetic field, to the only x-component of spin ($l_A \to 0$ and $l_R \to \infty$) corresponding to pure drift-diffusive transport as described by equations (12) and (14).

IOP Conf. Series: Journal of Physics: Conf. Series **1081** (2018) 012002 doi:10.1088/1742-6596/1081/1/012002

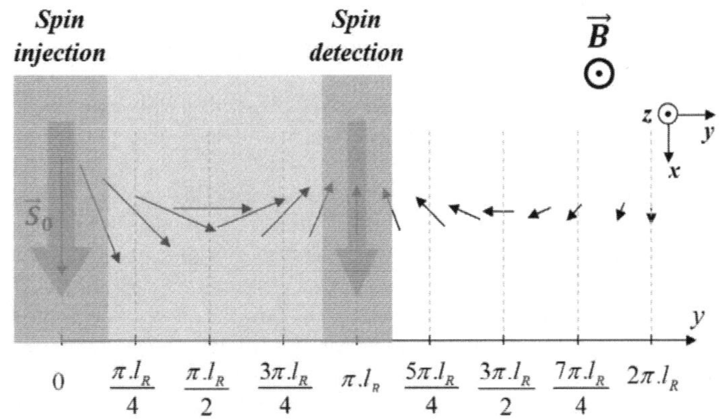

Figure 4. Top-view schematic of stationary state spin precession in the semiconductor channel under an external magnetic field along the z-direction.

According to equations (17) and (18), the characteristic decay length l_A and spatial period l_R depend on the applied magnetic field and the injected electron current. Figures 5 and 6 plot RT attenuation and oscillation lengths in n-silicon (with a dopant concentration of 10^{16} cm^{-3}) as functions of magnetic field and electrons injected density.

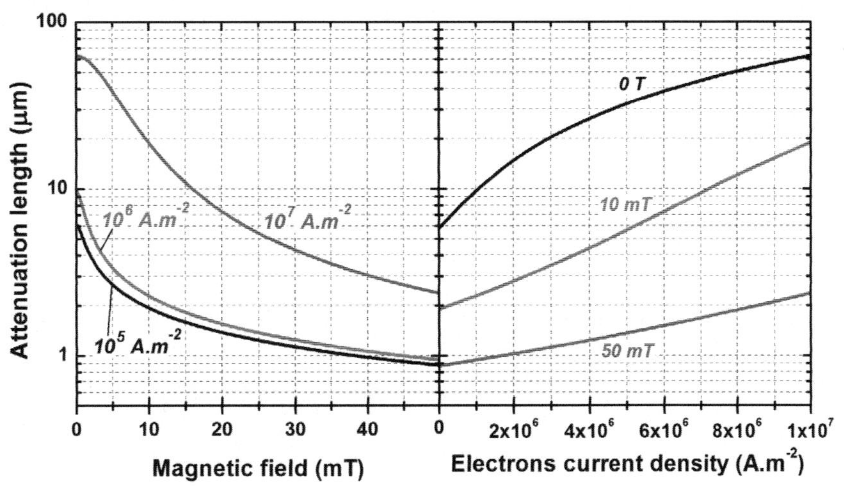

Figure 5. Effect of external magnetic field and injected electrons current density on RT attenuation length of the spin for low-doped n-silicon (n = 10^{16} cm^{-3}). Spin relaxation time is 10^{-8} s.

As shown in figure 5, the attenuation length l_A is strongly affected by the quite weak magnetic field (of several mT). This is because the electrons with the spin-up and spin-down orientation progress in semiconductor with different velocities (as seen in the drift-diffusion spin transport), which leads to

IOP Conf. Series: Journal of Physics: Conf. Series **1081** (2018) 012002 doi:10.1088/1742-6596/1081/1/012002

the spatial decoherence of spin precessions. For higher magnetic field, the strong contribution of the spin-magnetic field coupling compared to diffusion-drift competing mechanism ($\Omega \gg \frac{J^2}{qnD}$) results in the decreasing of the attenuation length. This result is consistent with equation (15) and clearly evidenced in the right graph of figure 5 showing the attenuation length as a function of injected current density for different values of applied magnetic field.

As documented in figure 6, the oscillation length decreases with the applied magnetic field, and is almost the same for the injected current densities of 10^5 A.m^{-2} and 10^6 A.m^{-2}. This result is very interesting from a practical point of view, since it shows the possibility to manipulate spins of low-doped silicon-based devices with different electrical characteristics (but operating in the above current density range) with the same magnetic field (that could be generated with a current line such in MRAMs). In addition, for weak magnetic field values, the linear trend of the log-log left graph of figure 6 indicates that the oscillation length is rather proportional to the inverse of weak applied magnetic field. Under strong magnetic fields, the spins process with increasing spatial frequencies, shortening then the spatial period of the spin oscillations.

Moreover, as shown in the right graph of figure 6 for different values of magnetic field, in contrast with the attenuation length, the spatial rotation period l_R is almost independent of the injected electron current density in the weak-injection range, and increases beyond.

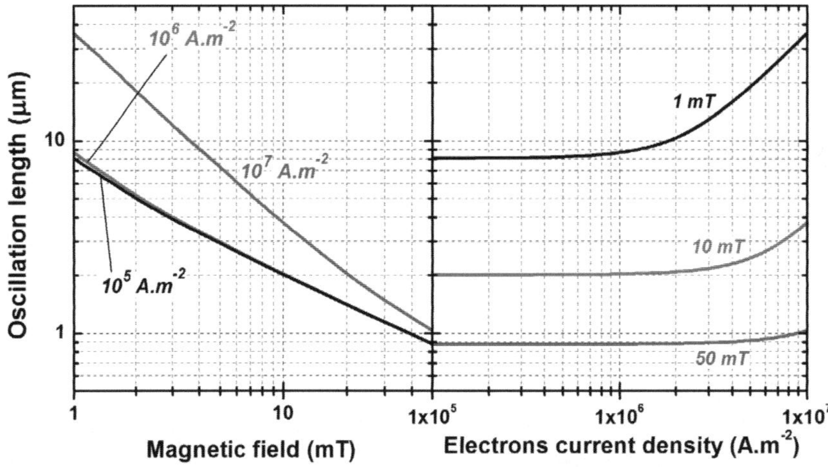

Figure 6. Effect of external magnetic field and injected electrons current density on RT oscillation length of the spin for low-doped n-silicon ($n = 10^{16}$ cm^{-3}). Spin relaxation time is 10^{-8} s.

Figure 7 depicts the spatial distribution of the normalized S_X spin component, in low-doped n-silicon ($n = 10^{16}$ cm^{-3}) at room temperature, for different values of applied magnetic field and injected current density.

As we can see in the left graph, under weak magnetic field, the spin oscillation can be clearly observed only for the relatively strong currents densities (10^7 A.m^{-2}), when the drift current dominates the diffusion one, increasing then the attenuation length (as shown in figure 5). For low current densities (10^6 A.m^{-2}), the spin oscillation is strongly damped and the effect of external magnetic field is reduced to a considerable reduction of the "effective" diffusion spin length. This effect is very strong and the external magnetic field of several milli-Tesla can completely depolarize electrons approaching the spin detector, about one micron far from the injector.

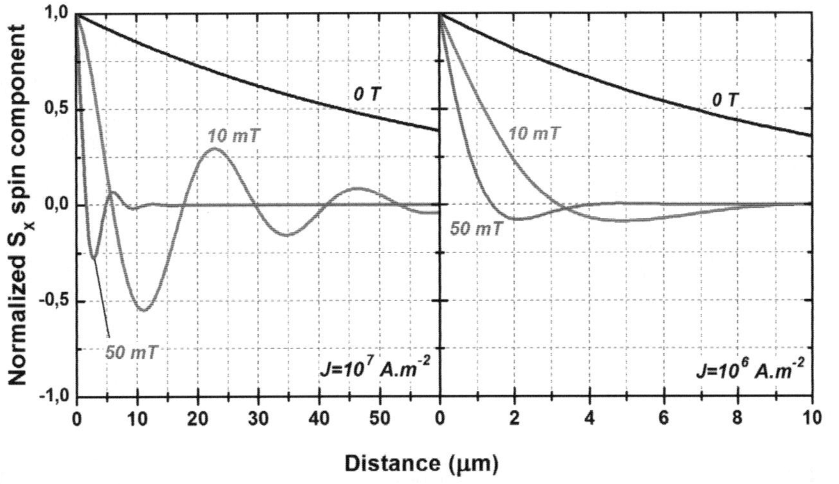

Figure 7. RT spatial distributions of the normalized S_X spin component, in low-doped n-type silicon ($n = 10^{16}$ cm^{-3}), for different values of transverse external magnetic field (0 T, 10 mT and 50 mT) and for injected current densities of 10^7 A.m^{-2} (left graph) and 10^6 A.m^{-2} (right graph). Spin relaxation time is 10^{-8} s.

3. Conclusion

We developed a drift-diffusive theoretical model of spin transport and spin precession under an external magnetic field in semiconductors, and we investigated carrier spin and current polarizations of electrons with spin-up and spin-down orientations. The developed analysis was used to make numerical evaluations for relatively low-doped silicon at room temperature.

It was found that the spin length in silicon is about several microns at room temperature and increases with the drift current. Furthermore, a rather weak external magnetic field (of several milliTesta) strongly affects spin transport and thus can be used for spin processing in silicon-based spintronic devices.

Apart of being the most used material in semiconductor industry, the silicon is then found to be a promising material for spintronic applications, where the spin transport and the spin manipulation along relatively long and different distances are necessary.

Acknowledgments
This work was supported by the French Research Network in Micro and Nano Technologies (RMNT), project N°122 MEMIS.

References

[1] Rochdi N, Tonneau D, Jandard F, Dallaporta H, Safarov V and Gautier J 2008 Electrical conductivity of ultra-thin silicon nanowires *J. Vac. Sci. Techno.* B **26** 159
[2] Rochdi N, Tonneau D, Jandard F, Dallaporta H and Safarov V 2008 Fabrication and electrical properties of ultra-thin silicon nanowires *Phys. Status Solidi* (a) **205** 1157
[3] Abed H, Sahaf H, Reguer A, Rochdi N, Tonneau D, Bedu F, Dallaporta H and Jamgotchian H 2010 Silicon nanowires synthesis on a submicronic terminal: Structural and electrical characterization *J. Appl. Phys.* **107** 024310

[4] Raïssi M, Regula G, Hadj Belgacem C, Rochdi N, Bozzo-Escoubas S, Coudreau C, Holländer B, M. Fnaiech, D'Avitaya F A and Lazzari J L 2011 Different architectures of relaxed $Si_{1-x}Ge_x$/Si pseudo-substrates grown by low-pressure chemical vapor deposition: Structural and morphological characteristics *J. Cryst. Growth* B **328** 18
[5] Gregg J, Allen W, Viart N, Kirschmann R, Schille C S J P, Gester M, Thomson S, Sparks P, DaCosta V, Ouandjela K and Skvarla M 1997 The art of spin electronics *J. Man. Magn. Mater.* **175** 1
[6] Zutic I, Fabian J and Sarma S D 2001 A proposal for a spin-polarized solar battery *Appl. Phys. Lett.* **79** 1558
[7] Dennis C L, Sirisathitkul C, Ensell G J, Gregg J F and Thompson S M 2003 High current gain silicon based spin transistor *J. Phys.* D **36** 81–87
[8] Dennis C L, Tiusan C V, Gregg J F, Ensell G J and Thompson S M 2005 Silicon spin diffusion transistor: materials, physics and device characteristics. *IEE Proc-Circuits Device Syst.* **152** 340
[9] Fabian J, Zutic I and Sarma S D 2004 Magnetic bipolar transistor *Appl. Phys. Lett.* **84** 85
[10] Zutic I, Fabian J and Sharma S D 2004 Spintronics: Fundamentals and applications *Rev. Mod. Phys.* **76** 323
[11] Datta S and Das B 1990 Electronic analog of the electro-optic modulator *Appl. Phys. Lett.* **56** 665
[12] Rashba E I 2000 Theory of electrical spin injection: Tunnel contacts as a solution of conductivity mismatch problem *Phys. Rev.* B **62** R16267
[13] Fert A and Jaffres H 2001 Conditions for efficient spin injection from a ferromagnetic metal into a semiconductor *Phys. Rev.* B **64** 184420
[14] Schmidt G, Ferrand D, Mollenkamp L W, Filip A T and Van Wees B J 2000 Fundamental obstacle for electrical spin injection from a ferromagnetic metal into diffusive semiconductor *Phys. Rev.* B **62** R4790
[15] Motsnyi V F, Safarov V I, Van Dorpe P, Das J, Van Roy W, Goovaerts E, Borghs G and De Boeck J 2003 Electrical spin injection in a ferromagnetic metal/insulator/semiconductor tunnel heterostructure *J. Supercond.* **16** 671
[16] Handbicki A T, Jonker B T, Itskos G, Kioseoglou G and Petrou A 2002 Efficient spin injection from magnetic metal/tunnel barrier contact into a semiconductor *Appl. Phys. Lett.* **80** 1240
[17] Motsnyi V F, Van Dorpe P, Van Roy W, Goovaerts E, Safarov V I, Borghs G and De Boeck J 2003 Optical investigation of electrical spin injection into semiconductors *J. Phys. Rev.* B **68** 245319
[18] Rochdi N, Raïssi M, Vizzini S, Coudreau C, Lazzari J L, Aufray B, Oughaddou H and Arnaud D'Avitaya F 2011 Structural and electrical characterizations of nanometer scaled aluminum oxide in metal / insulator / silicon (001) heterostructures *Glob. J. Phys. Chem.* **2** 230
[19] Raïssi M, Vizzini S, Langer G, Rochdi N, Oughaddou H, Coudreau C, Nitsche S, Arnaud D'Avitaya F, Aufray B and Lazzari J L 2010 Interfacial solid phase reactions in cobalt/aluminium-oxide/silicon(001) system *Thin Solid Films* **518** 5992
[20] Rochdi N, Liudvikouskaya K, Descoins M, Raïssi M, Coudreau C, Lazzari J L, Oughaddou H and Arnaud D'Avitaya F 2011 Surface morphology and structure of ultra-thin magnesium oxide grown on (100) silicon by atomic layer deposition oxidation *Thin Solid Films* **519** 6302
[21] Sarpi B, Rochdi N, Descoins M, Daineche R, Bertoglio M, Girardeaux C, Baronnet A, Perrin-Toinin J, Bocquet M, Djafari Rouhani M, Hemeryck A and Vizzini S 2015 Oxidation of Mg atomic monolayer onto silicon: A road toward MgO_x/Mg_2Si (11−1)/Si (100) heterostructure *Surf. Sci.* **642** L1
[22] Aronov A G and Pikus G E 1976 Spin injection into semiconductors *Sov. Phys. Semicond.* **10** 698

[23] Yu Z G and Flatté M E 2002 Electric field dependent spin diffusion and spin injection into semiconductors *Phys. Rev.* B **66** 201202; Yu Z G and Flatté M E 2002 Spin diffusion and injection in semiconductors. Electric field effects *Phys. Rev.* B **66** 235302

[24] Lépine D 1970 Spin resonance of localized and delocalized electrons in phosphorus doped Silicon *Phys. Rev.* B **2** 2429

[25] Rochdi N 2007 Proprietes de transport de microstructures et nanostructures de silicium *Ph.D. Thesis - University of Mediterranean* (Archives ouvertes HAL: tel-00192856) p81

IOP Conf. Series: Journal of Physics: Conf. Series **1081** (2018) 012003 doi:10.1088/1742-6596/1081/1/012003

Effect of crystallization time on the structure of the Aluminosilicate glass-ceramic

R Souag[1], N Kamel[2], D Moudir[2], Y Mouheb [2] and S Kamariz [2]

[1]URMPE, Research Unit for Materials, Processes and Environment, Faculty of Engineering Sciences, University M'hamed Bougara Boumerdes, 35000, Boumerdes,Algeria

[2]Nuclear Research Center of Algiers, Nuclear Technology Division, 2. Bd Frantz Fanon, BP 399, RP-Algiers.

E-mail: souagrafika@yahoo.fr

Abstract. All This study deals with the influence of the time of crystallization on the structure of an aluminosilicate glass-ceramic ceramized by a nucleation –crystallization treatment at 790°C for 2 h, and 900°C, for different periods of time ranging from 2 to 12 h. For the whole of the materials, Archimedes density is between 2441-2578 kg/m^3. Both X-ray diffraction and scanning electron microscopy analyses reveal two main crystalline phases for the whole of the heating treatments, namely spodumen and leucite. These phases grow regularly with the crystallization time. In addition, aqueous stability testing was carried out using the standard MCC-2 static leach test method at 90 °C and V(ml) = 30 S(cm^2) for a series of five test durations - 7, 14, 28, 35 and 42 days. The X-ray fluorescence analytical technique was used to determine the concentration of lanthanum (La) and molybdenum (Mo) leached in our material. The dissolution parameters, in this case the normalized elemental mass loss (NL) and the leaching rate (τ) of each element, were determined from these concentrations. The results demonstrate that The leaching tests showed that the material is stable and it stabilizes at low values (2.5 10^{-9} Kg/m^2) for La when the ceramization time does not exceed 4h beyond this, Leaching equilibrium and the passivation phase is broken. For Mo and for all glass ceramics at different times of crystallization, the equilibrium reaches (3.15 10^{-8} Kg/m^{2}j) and the passivation layer is formed. The glass ceramics at different time of cristallization presently studied appears to have a good chemical durability.

1. Introduction

Glass-ceramics (GC) containing highly durable crystalline particles homogeneously dispersed in the bulk of a glass matrix have been proposed as important candidates for the immobilization of high level radioactive wastes (HLW). They show potential interest to the confinement of radioactive waste by presenting a double shell protection, the first envelope is the crystalline phase, which traps radionuclides, and the second one is the glass as a second barrier surrounding the crystal [1]. In this context, glass-ceramics matrices constituted by crystals distributed homogeneously within a glass matrix and exhibiting increased performances (chemical durability and capacity of reception) compared with the aluminoborosilicate glasses currently used to confine all the waste are particularly interesting [2-5]. This study deals with the influence of the time of crystallization on the structure of an aluminosilicate glass-ceramic ceramized by a nucleation –crystallization treatment at 790°C for 2 h, and 900°C, for different

IOP Conf. Series: Journal of Physics: Conf. Series **1081** (2018) 012003 doi:10.1088/1742-6596/1081/1/012003

periods of time ranging from 2 to 12 h. The influence of the time of crystallization (t_c) on the nature of the crystalline phases formed in the bulk of the material is studied. In addition, aqueous stability testing was carried out using the standard MCC-2 static leach test method at 90 °C.

2. Experimental

The GC chemical composition is given in table 1. The following commercial reagents are used for the synthesis: Al_2O_3 (FLUKA), Fe_2O_3 (MERCK, Purity ≥99%), Na_2O (Merck, purity≥99.5%), K_2O (MERCK, Purity≥99%), Li_2O (MERCK, Purity ≥ 99%), MgO (FLUKA, Purity ≥ 97%), MoO_3 (MERCK, Purity≥99.5%), SiO_2 (Prolabo), ZrO_2 (Aldrich, Purity≥ 99%) and Y_2O_3 (MERCK, Purity≥99%). The rare earth oxides are dried at 1000 °C, and the other oxides at 400 °C overnight. This step is important to ensure homogeneity of synthetic products, and obtain materials with isotropic properties.

La_2O_3 is prepared by calcination at 450 °C from La $(NO_3)_3$ 6 H_2O (Fluka, purity≥99.99%). All reagents are finely ground in a manual agate mortar to obtain a particle size of about 20 μm before the preparation of the mixtures.

Table 1. Chemical composition (in molar %.) of the based glass per 100 g of mixture.

Oxides	ZrO_2	Al_2O_3	MgO	Na_2O	Fe_2O_3	K_2O	Li_2O	La_2O_3	Y_2O_3	MoO_3	SiO_2	Totale
Molar %	4.5	15.0	5.0	6.0	2.0	2.9	12.0	0.5	0.5	0.5	48.6	100.0

The powder mixtures are homogenized in an Automatic Sieve Shaker D403 shaker for 6 h, to ensure good dispersion of the mixture. Both nucleation and crystallization temperatures are deducted from trace DTA (differential thermal analysis) analysis of the parent glass. Nucleation is carried out in the same furnace as that used for the glass melting of the oxide mixture at a nucleation temperature (Tn) such as: Tn = Tg + 20, thus 790 °C, for 2 h. Grain growth can be performed at a temperature close to the crystallization temperature of the base glass, Tc, which is 900 °C. The crystallization time (t_c) is optimized as following: 6, 9 and 12 h. The GCs are cooled in open air to room temperature. The powders densities (*ρ_A*) are measured by the Archimedes method. XRD analysis of the materials is performed by a Philips X'Pert Pro diffractometer, with Cu Kα1 ray (λ= 1.5418 Å), using Philips X'Pert High Score software for phase identification [6].

A scanning electron microscope (SEM) Philips XL30 allowed micrographic observations. Aqueous stability testing was carried out using the standard MCC-2 static leach test method at 90 °C and V(ml) = 30 S(cm^2) for a series of five test durations - 7, 14, 28, 35 and 42 days.

The X-ray fluorescence (XRF) technique is used to measure the Lanthane (La) and Molybdene (Mo) dissolution rate in all the aqueous leachates. The employed XRF spectrometer is a Magix-Pro PAN Analytical equipment.

The normalized elemental mass loss in this solution (NL) is defined by the following relationship:

$$NL = Ci.Vo/So.Fi \qquad (1)$$

With:

C_i: is the total concentration of La or Mo leached ($kg.m^{-3}$),

Vo : the total volume of the leaching mixture (m^3),

So: the sample initial surface in actual contact with the leachate (m^2),

Fi: the La or Mo weight fraction in the material,

And the normalized leaching rate is expressed by the relation:

$$\tau i = \left(\frac{1}{Fi} + So\right) * \left(\frac{dmi}{\Delta t}\right) \quad (2)$$

With:
τi : the normalized leaching rate of la or Mo, expressed by $kg.m^{-2}.j^{-1}$
mi: the leached mass of La or Mo during a time interval Δt expressed in kg,
Δt : is the interval time leach (j),

3. Results and Discussions

The density of the obtained GCs with different crystallization times (t_c) is measured by the Archimedes method. The results are given in Table 2.

Table 2. Variation of the GC densities (kg/m^3) according to crystallization time (h).

Cristallisation time t_c (h)	2	4	6	9	12
Density ρ_A (kg/m^3)	2441	2492	2509	2532	2573

The GC density increases with the crystallization time. It is between 2441 and 2573 kg/m^3 for all the materials. This density rise can be attributed to the gradual germination and continuous heavy crystalline phases as a function of crystallization time. However, this hypothesis must be confirmed by further investigations of the GCs. Seokju J. and al. [7] report densities values close to our results ($\approx$ 2.8 g/cm^3) for cordierite glass-ceramics synthesized by a melting at 1450°C and a crystallization and crystal growth treatments, at 750 and 1020°C, respectively.

The GC XRD analysis gave the diffractograms depicted in Figure 1. The results of the phases identification are shown in Table 3.

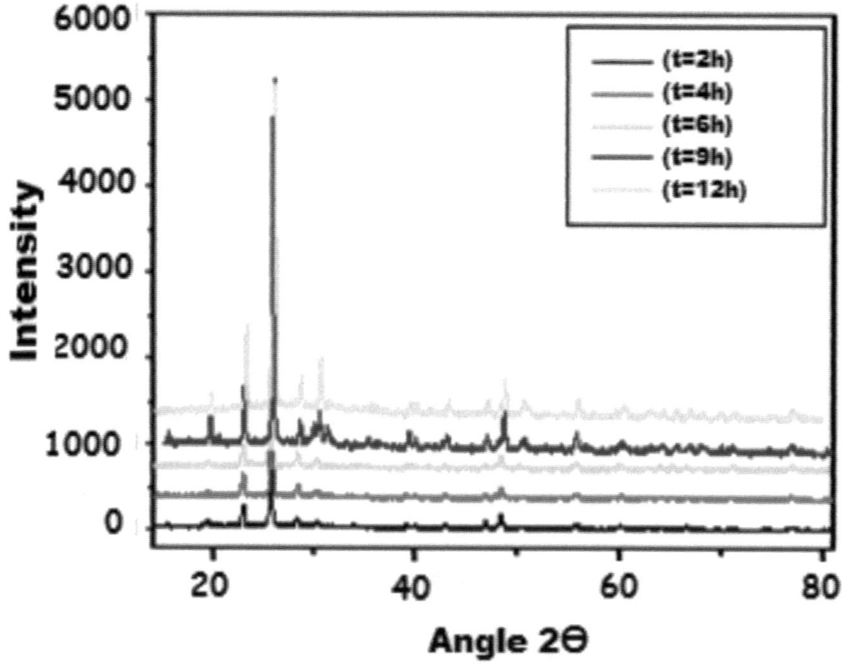

Figure 1. Diffractograms of the GCs for different crystallization times.

We find that for all GCs and at different crystallization times, a major phase of the aluminosilicate type of the pyroxene (Si_2O_6) family, with chemical formula $MAlSi_2O_6$ (M = Li, K, Na or Mg) has been identified. This phase is known as a containment barrier against the dissipation of radionuclides, the rate of the formation of this phase increases as with crystallization time increases. At t_c = 2 h, this aluminosilicate phase of spodumene nature has already formed at 39%. Two major phases were observed and distributed equally at t_c = 4h, these two phases being a spodumene aluminosilicate with 48% and a molybdate phase with a germination rate of 49% (15% $Sm_2(MoO_4)_3$+ 34% $Fe_2(MoO_4)_3$). At t_c = 9h, a main aluminosilicate phase of leucite ($KAlSi_2O_6$) is assigned to the crystallized GC.
This aluminosilicate phase of the pyroxene family crystallizes progressively with the crystallization time to the detriment of the molybdate phase ($NaLa(MoO_4)_2$), which reaches 5% at t_c = 12 h. The highest aluminosilicate content, but doped in two different ways, is obtained for the maximum crystallization temperature (t_c = 12h). It exceeds 92% (60 % $Mg_{0.6}Al_{1.2}Si_{1.8}O_6$ and 32 % $NaAlSi_2O_6$).
Therefore, one can infer that the rise of crystallization time leads to the preferential germination of aluminosilicate phases.

Table 3. The GCs crystalline phases identification for different crystallization times.

Crystallization time (t_c) (h)	Chemical composition	JCPDS data [2]
2	39 % LiAlSi2O6	(01-074-1106)
	48% CaAl4O7	(01-076-0706)
	1 % ZrO2	(01-079-1770)
	12% $Na_{0.048}(Ca_{1.53}Y_{0.47})(Al_{0.48}Be_{2.52}Si_{12}O_{30}$	(01-080-2240)
4	48% LiAlSi2O6	(01-074-1106)
	3% LiFeO2	(01-089-2783)
	15% Sm2 (Mo O4)3	(00-024-1000)
	34% Fe2 (Mo O4)3	(01-072-0935)
6	79 % $LiAlSi_2O_6$	(01-074-1106)
	19% $Pr_2Zr_3(MoO_4)_9$	(00-051-1851)
	2% $Zr_{0.935}Y_{0.065}O_{1.968}$	(01-080-2187)
9	82 % $KAlSi_2O_6$	(01-081-2221)
	11 % $(K_{2.5}Na_{0.5})Na(MoO_4)_2$	(01-088-0302)
	7 % $Mg1_{.55}Fe_{1.6}O_4$	01-080-0073)
12	60 % $Mg_{0.6}Al_{1.2}Si_{1.8}O_6$	(01-075-1568)
	32 % $NaAlSi_2O_6$	(01-071-1504)
	5 %$NaLa(MoO_4)_2$	(00-024-1103)
	3 % $Gd_3Fe_2Fe_3O_{12}$	(01-072-0141)

The Observation of the microstructure of the studied GCs is carried out by scanning electron microscopy (SEM). The microscope used in this study is Philips ESEM XL 30 equipment. The micrographs are shown in Figures 2.a, 2.b, and 2.c. These micrographs allowed us observing both types of phases: the crystalline and glassy ones, and thus confirm the presence of the main crystalline phases, identified by XRD analysis in the ceramic.

2.a **2.b** **2.c**

Figure 2. SEM micrographs of the GCs for different Tc values.

The variation of the normalized elemental mass loss (N_L) and the leaching rate of La and Mo (τ_i) are calculated using the formulas (1) and (2) of section 2, respectively.
The results of the evolution of N_L versus time are shown in figure 3, for all the tests.

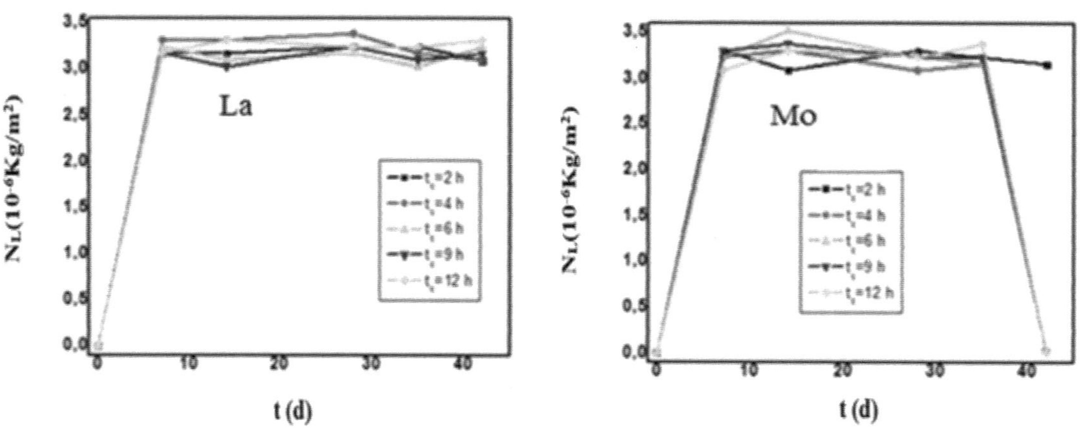

Figure 3. Evolution of the normalized mass loss (10^{-6}Kg/m^2) of La and Mo in leachates, function of time and crystallization time of the GCs.

For all GCs, the normalized mass loss stabilized around 3.10 10^{-6}Kg /m^2 and 0.4$10^{-6}$Kg/m^2 for La and Mo, respectively; whereas the Mo content is greater than that of La in the base glass, the latter is more releasable than Mo. This proves that lanthanum is probably bound to the glass by weak bonds. It is this fraction which is leached, compared to the small amount of La which is found in the crystalline phases. Especially since the quantities released are larger with the crystallization time (period when a molybdate phase doped with lanthanum appears). The Mo is released with a low content because it is confined in the molybdate phase which appears for all the crystallization times and confirms our results of DRX.
Yong He et al. [8] studied the leaching (MCC1) of an apatite-rich glass-ceramic with a crystallization temperature of 825 and 875 ° C for 2 h. They report N_L (2.7 10^{-6} Kg/m^2) values for Mo greater than our results obtained for the MCC2 test (0.41 10^{-6} Kg/m^2). This shows that we have a better result despite the aggressive environment we used (90 ° C). For the lanthanum actinide simulator, the value of the

normalized mass loss is 3.10 10^{-6} Kg/m^2, which is significantly higher than that found by our previous results (0.4 10^{-6} Kg/m^2) for the leaching of Ce determined from Zr-glass-ceramic (T_c = 1010 ° C and t_c= 2h) [9].
The results of the evolution of the La and Mo leaching rate (τ) as a function of time and cristallization time of GCs are gathered in figure 4.

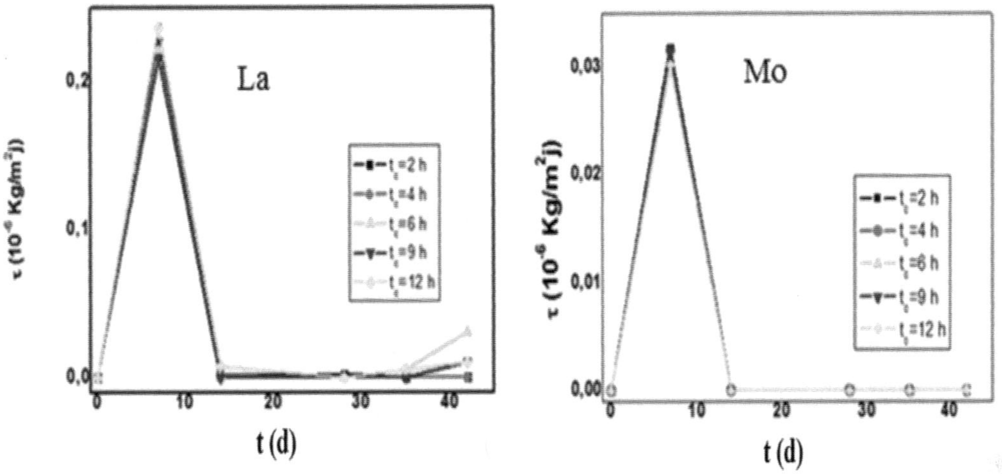

Figure 4. Evolution of the leaching rate ((10^{-6} Kg/m^{2}j)) of La and Mo in leachates, function of time and crystallization time of the GCs

The leaching rate in La and Mo increases rapidly to reach a maximum on seventh day and then decreases and stabilizes at low values (25 10^{-10} Kg / m^{2}d) for La when the ceramization time does not exceed 4 h. Beyond this, the matrix continues to dissolve for times greater than 35 days, showing that the La is leachable, and therefore is not protected by the spodumene and molybdate ceramics formed and as a result it is dissolved in the glass-ceramics and is weakly Bound to the glass and probably close to the surface thereof. Thus, for a time of ceramization greater than 4 h, there is no leaching equilibrium and the passivation phase is broken beyond 35 days of leaching for La. For Molybdenum and for all Glass-ceramics with different crystallization times, the equilibrium is reached and the passivation layer is formed.
Stéphane Gavarini et al. [10] also carried out a soxhlet dissolution test at pH = 5 of a La and Ce-rich aluminosilicate glass. They report dissolution rates of (3.5-6.9) 10^{-10} Kg / m^2 of La and Ce respectively, after one month (30 days) of testing. These values are very small compared to our values (25 10^{-10} Kg / m^{2}d) due may be to the acid medium used by the authors.
As a conclusion, the leaching equilibrium is reached for all the glass-ceramics with a crystallization time greater than 4 h except for the Lanthanum or the passivation phase is broken beyond 35 days of leaching. This shows that our glass-ceramics are durable, and can be recommended as actinide confinement materials.

4. Conclusion
In this study, we performed the synthesis and characterization of glass-ceramic matrices based on an aluminosilicate glass, able to incorporate elements of radioactive waste (fission products and lanthanides) in their structure. The employed synthesis method is the devitrification of a parent glass. It is carried out in several stages: a double fusion of the oxides mixture at 1350 °C, a nucleation at 790 °C, and a crystal growth at 900 °C, for different periods of time ranging from 2 to 12 h. Both X-ray diffraction and scanning electron microscopy analyses reveal two main crystalline phases for the whole of heating treatments, namely spodumen and leucite. These phases grow regularly with the crystallization time. In addition, aqueous stability testing was carried out using the standard MCC-2

static leach test method at 90 °C. The results demonstrate that The leaching tests showed that the material is stable and it stabilizes at low values (2.5 10^{-9} Kg/m^2) for La when the ceramization time does not exceed 4h beyond this, Leaching equilibrium and the passivation phase is broken. For Mo and for all glass ceramics at different times of crystallization, the equilibrium is reached (3.15 10^{-8} Kg/m^2j) and the passivation layer is formed. The glass ceramics at different time of cristallization presently studied appears to have a good chemical durability. The studied glass-ceramics have satisfactory mineralogical compositions, because they contain suitable crystalline phases for the double confinement of actinides, such as aluminosilicates (spodumene) and molybdate.

5. References

[1] Caurant D, 2009 *Nova Science Publishers*, ISBN 978-1-60456-174-6.
[2] Maddrell E. Thornber S, Hyatt N C 2015 *J. Nucl. Mater.* **456** 461
[3] Palanivel R and Velraj G 2007 *INDIAN J. PURE & APPL PHYS.* June **45**, 501
[4] Kaur R, Singh S and Prakash P O, Physica 2012 *Physica* B **407** 4765
[5] Petrovic R, Janackovic D, Zec S, Drmanic S and Kostic-Gvozdenovic L J 2003 *J. Sol-Gel Sci. Technol* **28** 111
[6] JCPDS data Philips X'Pert High Score Package, 2004, *International Center for Diffraction Data*, Newtown Square PA.
[7] Seokju J and Seunggu K 2012 *Ceram. Int.* **38S** S543
[8] Yong H, Weimin B, Chongli S 2002 *J. Nucl. Mater.* **305** 202.
[9] Souag R 2015 Thèse de doctorat, Université M'Hamed Bougara, Boumerdès, Algeria
[10] Gavarini S 2003 Thèse de doctorat. Orléans. France

Structural and optical properties of Yb doped ZnO nanorods synthesized via a green process

M Achehboune[1,3,*], M Khenfouch[2,3], I Boukhoubza[1,3], B Mothudi[2], I Zorkani[1] and A Jorio[1]

[1]University Sidi Mohammed Ben Abdellah, Faculty of Sciences Dhar el Mahraz, Laboratory of Solid state Physics, Group of Nanomaterials and Renewable Energies, PO Box 1796 Atlas Fez 30 000, Morocco

[2]UNISA university of South Africa, Department of Physics, College of Science, Engineering and Technology, Science Campus, Cnr Christiaan de Wet & Pioneer Avenue Florida 1709, Johannesburg, South Africa

[3]Africa Graphene Center, Physics department, Eureka building, College of Science, Engineering and Technology, Science Campus, University of South Africa, Cnr Christiaan de Wet & Pioneer Avenue Florida 1709, Johannesburg, South Africa.

* E-mail: achehboune.mohamed01@gmail.com

Abstract. Recently, ZnO Nanostructures have attracted considerable attention due to their various morphologies, easy synthesis and excellent physical properties for fabricating optoelectronic devices. In this study, we report the synthesis of Yb doped ZnO nanorods via a green process using plant extracts. Their morphology, structural and optical properties were characterized by different techniques. The obtained materials show promising properties and remarkable effects of the dopant. Our investigations and discussions were conducted to reveal our synthesized materials ability for optoelectronic applications such as solar cell and light emitting diodes devices.

1. Introduction

ZnO is a natural n-type semiconductor with a direct wide band gap (3.37 eV) and large exciton binding energy (60meV). Its optical and electrical properties make it useful as an attractive and promising material for optoelectronic and photonic applications such as solar cells, lasers, light emitting diodes, photodetectors [1-2]. In addition, ZnO is a material easily synthesized under several morphologies such as quantum dots [3], nanorods [4], nanowires [5], nanobelts [6], nanoflowers [7], etc. So far, various methods have been used to synthesis ZnO nanostructures, such methods include green synthesis [8], hydrothermal [9], sol-gel [10], chemical vapor deposition [11] and microwave assisted [12]. The green synthesis of ZnO nanorods makes use of economical and safe reagents, non-toxic, environmental friendly methods that have attracted much attention recently.

Structural and optical properties of ZnO nanostructures can be modified and improved by doping/co-doping processes using metallic dopants such as Al [13], Ga [14], Na [15], etc. In recent years rare earth doped ZnO nanostructures have attracted the attention of scientist researches since they are useful and

IOP Conf. Series: Journal of Physics: Conf. Series **1081** (2018) 012004 doi:10.1088/1742-6596/1081/1/012004

better luminescent materials owing to their 4f intra shell transitions [16]. In the literature, studies of rare earth doped ZnO are scarce, we find among them for example, the study conducted by Senol [17] which reported the synthesis of rare earth (Er-Yb) doped ZnO nanostructures using the hydrothermal method. Zamiri et al [18], have also synthesized rare earth (Er, La, Yb) doped ZnO nanostructures through wet precipitation method and studied their structural and optical properties. Soumahoro et al [19] have reported the structural, optical and electrical properties of Yb-doped ZnO thin films synthesized by the spray pyrolysis method in order to investigate the effect of Yb ion in the ZnO matrix and the related optical properties of the films. Kumar et al [20] have also prepared Nd-doped ZnO thin films deposited on a glass substrate by sol-gel.

In our work, we report the structural and optical properties of pure and Yb doped ZnO nanorods; samples have been synthesized by hydrothermal method for pure ZnO and green hydrothermal method for Yb doped ZnO and their structural and optical properties were studied by Scanning Electron Microscopy (SEM), X-Ray Diffraction (XRD) and Photoluminescence (PL). The aim of this study was to investigate the effect of Yb dopant on the optoelectronic properties of ZnO nanorods.

2. Experimental details

ZnO precursors used in this work were purchased from Sigma-Aldrich. To synthesize ZnO nanorods, an equi-molar aqueous solution of 0.1 M zinc nitrate ($Zn(NO_3)_2$) and hexamethylenetetramine ($C_6H_{12}N_4$, HMT) was prepared using deionized water. This mixture was heated at a constant temperature of 90°C in an oven for 24 h. The doping process followed the same steps with the addition of 10% of ytterbium chloride and 10 ml of tea leaves extract. Dry green tea leaves were purchased from Lipton company. In this setup, 0.1 g of these leaves were boiled at 70°C in a volume of 200 ml of de-ionized water for 40min. A brown solution of in. brown colored extract of pH=5 was obtained. The brown solution was filtered 3 times to eliminate residual solids. Determined volumes were token from this extract rich in polyphenols and used in the synthesis of our nanomaterials.

The obtained materials were deposited on Si (100) substrates and dried at 50°C for 2 hours in air.

Results and discussion

2.1. X-ray diffraction

Figure 1 shows the X-ray diffraction pattern of pure and Yb-doped ZnO nanorods. As can be seen from the spectra, the analysis of diffraction peaks of both pure and Yb doped ZnO reveals the presence of hexagonal wurtzite structure without any diffraction peaks of Yb related impurities. Results showed that the intensity of (002) peak and (101) peak decreases with the addition of ytterbium, while the intensity of (100) peak does not change; it means that the growth in the case of pure ZnO was in the three principal directions (reflections (100), (002), (101)), although for Yb-doped ZnO the direction (100) is the preferred one for growth (because the addition of Yb causes notable decrease in both (002) and (101) peak intensities), the growth in directions (002) and (101) is limited after doping which means that doping has an effect on crystallinity.

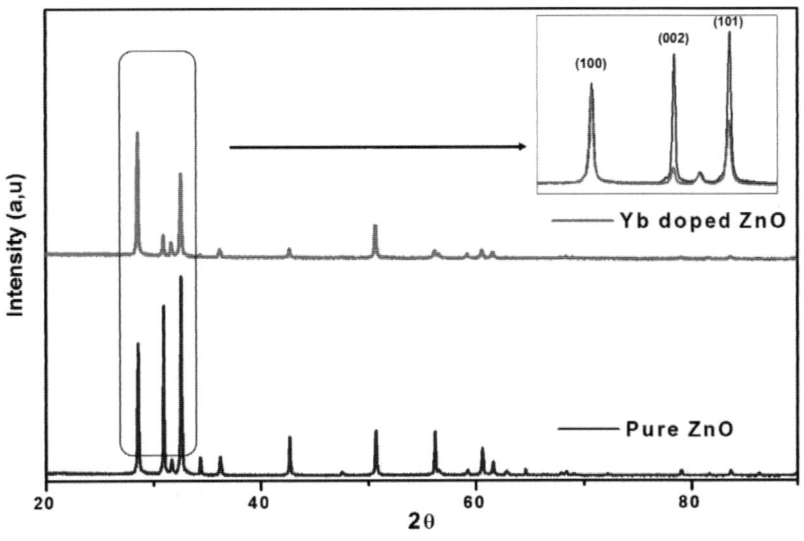

Figure 1. X-ray diffraction spectrum of pure and Yb doped ZnO nanorods.

2.2. Scanning electron microscopy

Morphological characterization of pure and Yb doped ZnO nanorods was performed by SEM observation, Figure 2 shows typical SEM image of ZnO nanorods prepared via a green process. As clearly shown in the Figure 2(a), the ZnO nanorods have an hexagonal shape, the average diameter and the length are about 500 nm and several micrometres (~4µm) respectively. ZnO nanorods were relatively grown vertically with a coexistence of two different sizes and high density (an average of 11 nanorods/ 10µm2). The graph at the top of the figure presents the distribution of nanorods taken from different areas of 10 µm2. The image of Yb doped ZnO in the Figure 2(b) indicates the formation of a bulk material with rarely located nanorods with different sizes in the edges, which means that both the nucleation and the growth were affected due to the dopant.

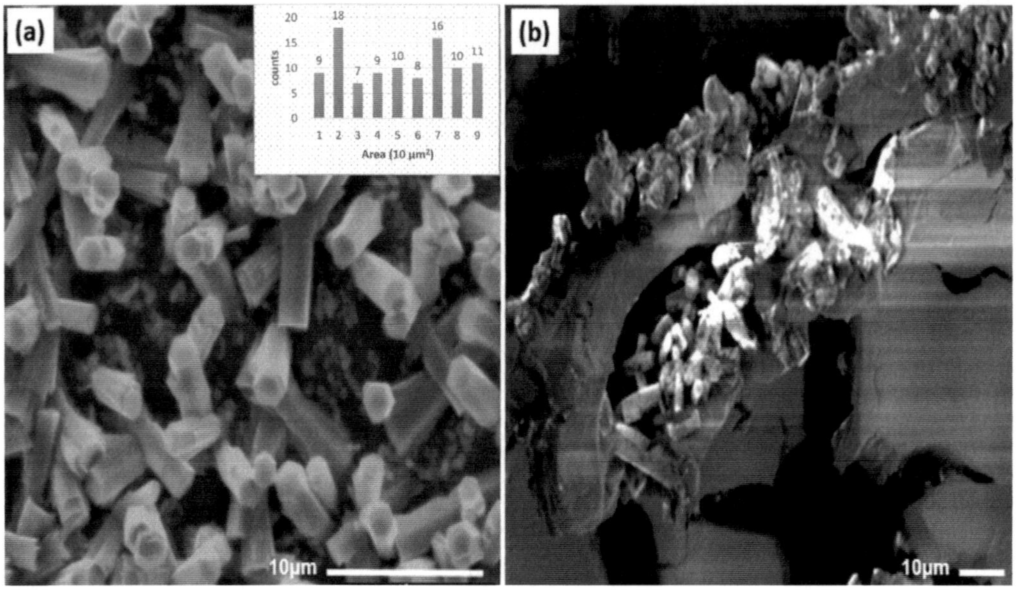

Figure 2. SEM images of pure and Yb doped ZnO nanorods.

2.3. Photoluminescence

Room temperature photoluminescence study has been performed for investigating the optical properties and the effect of Yb doping ZnO. It was known that the luminescence of ZnO largely depends on the synthesis method and most importantly the defects concentration in the bulk and on the surface of material. Figure 3 shows the obtained PL spectra of pure and Yb doped ZnO with 10% concentration of doping prepared respectively by hydrothermal and green hydrothermal method. It is clearly remarkable that pure ZnO peak is more intense than that Yb doped ZnO. A number of hypotheses have been proposed to explain this strong decrease in intensity, such as an increase in non-radiative recombination process [21], a higher concentration of Yb (10%) (because when the doping exceeds a given concentration the intensity decreases in the visible region [22]), the compensation of ions introduced by the Yb doping into the ZnO lattice or to bioactive molecules of tea introduced during the preparation which occupy the intrinsic defect of ZnO.

To better understand the nature and the origin of emissions, Figure 4 illustrates the normalized spectra of photoluminescence. The two samples exhibit two peaks in the visible region; the peak of pure ZnO was observed approximately at 560 nm (yellow emission). In general, oxygen vacancy (V_o) is the most widely accepted mechanism for the visible emission of ZnO; thus, the peak of pure ZnO is usually attributed to the doubly ionized oxygen vacancy ($V_o^{\bullet\bullet}$) [23]. In the PL spectra of Yb doped ZnO nanorods, the peak is approximately located at 545 nm (green and yellow emission), which can be attributed to the decrease of number of oxygen vacancies and interstitial oxygen and this emission comes from the recombination of the holes and the interstitial oxygen ions when electrons occupy the singly ionized oxygen vacancy ($V_o^{\bullet}$) [24]. Therefore, after doping with ytterbium, there is a slight shift of the observed peak to high energies; this effect could be attributed to the decrease of structures defects which were responsible of the emission in this region (520-620), this decrease may be due to the occupation of the Yb atoms, i.e. the creation of a level which limits the emission in this region.

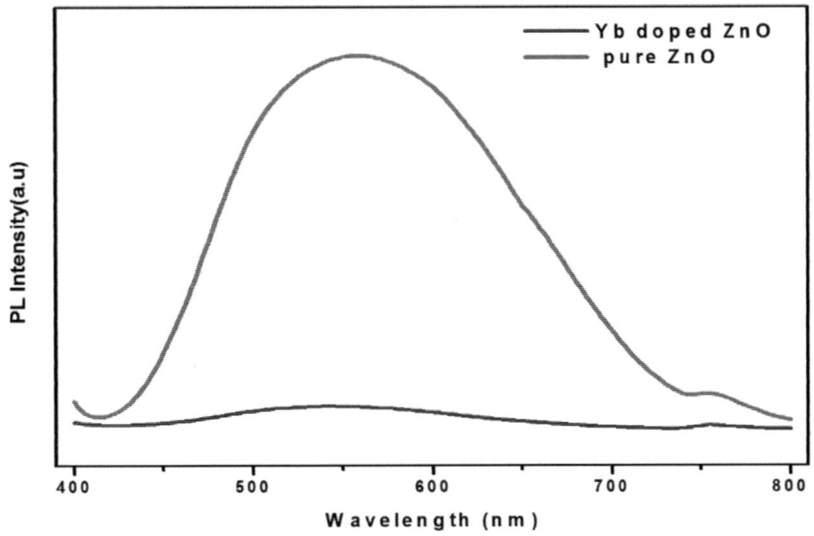

Figure 3. PL Spectra of pure and Yb doped ZnO Nanorods prepared respectively by hydrothermal and green hydrothermal method.

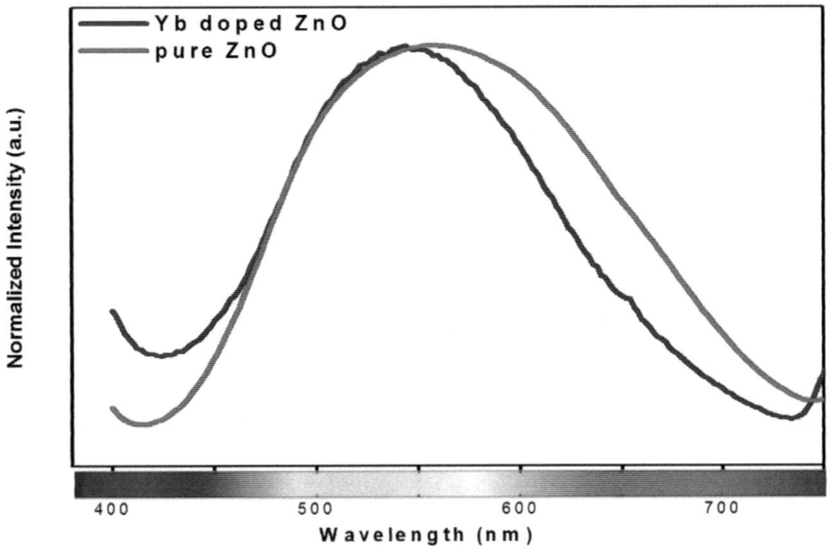

Figure 4. Normalized PL Spectra of Pure and Yb doped ZnO Nanorods.

Conclusion

This present study is focused on the study of structural and optical properties of Yb doped ZnO nanorods able to enhance their performance in the optoelectronic devices. Pure and Yb doped ZnO nanorods were prepared respectively via hydrothermal and green hydrothermal methods; the Yb concentration of doped

ZnO nanorods was 10%. Structural, morphological and optical properties have been studied, and the results revealed that both pure and Yb doped ZnO have the hexagonal wurtzite structure, with (100) as a preferred orientation for the growth of the Yb-doped ZnO. The doping conducted to a bulk material with rarely located nanorods with different sizes. The PL spectra show a strong decrease in intensity after doping with Yb and a slight shift to high energies.

Acknowledgements
This work was supported by the Faculty of Sciences Dhar el Mahraz USMBA (Morocco), University of South Africa Department of Physics (South Africa) and Africa Graphene Center (AGC).

References

[1] Ozgur U, Alivov Y I, Liu C, Teke A, Reshchikov M A, Dogan S, Avrutin V, Cho S J and Morkoç H 2005 *J. Appl. Phys.* **98** 041301
[2] Lukas S M and Judith L M D 2007 *MaterialsToday* **10** 40-48
[3] Qiao B, Suling Z, Zheng X and Xurong X 2016 *Chin. Phys. B* **25** 098102
[4] Cakir A C and Erten-Ela S 2012 *Advanced Powder Technology* **23** 655–660
[5] AlSalhi M S, Atif M, Ansari A A, Khun K, Ibupoto Z H and Willander M 2013 *Laser Phys.* **23** 065602
[6] Bhatti H S, Gupta A, Verma N K and Kumar S 2006 *J. Mater. Sci: Mater Electron* **17** 281–285
[7] Xu Y, Jin I, Li X, Han Y, Meng H, Wang T and Zhang X 2016 *Materials Research Bulletin* **76** 235–239
[8] Davar F, Majedi A and Mirzaei A 2015 *J. Am. Ceram. Soc.* **98** 1739–1746
[9] Edalati K, Shakiba A, Vahdati-Khaki J and Zebarjad S M 2016 *Materials Research Bulletin* **74** 374–379
[10] Chen Z, Li X X, Du G, Chen N and Suen A 2011 *Journal of Luminescence* **131** 2072–2077
[11] Cheng A J, Tzeng Y, Zhou Y, Park M, Wu T H, Shannon C, Wang D and Lee W 2008 *Appl. Phys. Lett.* **92** 092113
[12] Hasanpoora M, Aliofkhazraeia M and Delavari H 2015 *Procedia Materials Science* **11** 320 – 325
[13] Nur S R, Zainovia L, Azlan A A and Khairunisak A R 2016 *Materials Science Forum* **846** 459-464
[14] Lim S K, Hong S H, Hwang S H, Kim S and Park H 2013 *J. Mater. Sci. Technol.* **29** 39-43
[15] Ye Z, Wang T,Wu S, Ji X and Zhang Q 2017 *Journal of Alloys and Compounds* **690** 189–194
[16] Petersen J, Brimont C, Gallart M, Schmerber G, Gilliot P, Ulhaq-Bouillet C , Rehspringer J L, Colis S, Becker C, Slaoui A and Dinia A 2010 *J. Appl. Phys.* **107** 123522
[17] Senol S D 2016 *J Mater Sci: Mater Electron* **27** 7767–7775
[18] Zamiri R, Lemosa A F, Rebloa A, Ahangarb H A and Ferreira J M F 2014 *Ceramics International* **40** 523–529
[19] Soumahoro I, Schmerber G, Douayar A, Colis S, Abd-Lefdil M, Hassanain N, Berrada A, Muller D, Slaoui A, Rinnert H and Dinia A 2011 *J. Appl. Phys.* **109** 033708
[20] Poongodia G, Kumarb R M and Jayavelc R 2015 *Ceramics International* **41** 4169–4175
[21] Tomakin M 2012 *Superlattices and Microstructures* **51** 372–380
[22] Shrisha B V, Bhata S, Kushavahb D and Naik K G 2016 *Materials Today: Proceedings* **3** 1693–1701
[23] Kim S, Kim M S and Nam G 2012 *Electronic Materials Letters* **8** 445-450
[24] Wu X L, Siu G G, Fu C L and Ong H C 2001 *Appl. Phys. Lett.* **78** 2285

First steps of silicene growth on Ag(111)

Mohamed Rachid Tchalala[1], Hanna Enriquez[1], Andrew J. Mayne[1], Abdelkader Kara[2], Gérald Dujardin[1], and Hamid Oughaddou[1,3,*]

[1] Institut des Sciences Moléculaires d'Orsay, ISMO-CNRS, Bât. 210, Université Paris-Sud, F-91405 Orsay, France
[2]Department of Physics, University of Central Florida, Orlando, FL 32816, USA
[3] Département de Physique, Université de Cergy-Pontoise, F-95031 Cergy-Pontoise Cedex, France

**e-mail : Hamid.Oughaddou@u-psud.fr*

Abstract. In this paper we report on the first steps of silicene growth on Ag(111) using scanning tunneling microscopy. We show that the topmost atomic layer is composed of both silicon and silver. The STM observations are consistent with an exchange process between the silicon and silver atoms preferentially taking place at the step edges of the Ag substrate. In addition, silicon stripes are observed as precursors of the formation of the silicene sheet.

1. Introduction

Two-dimensional (2D) materials hold great promise for future electronic devices [1]. Indeed, even though graphene is now used in a number of applications, its semi-metal character and the absence of a band gap [2] have posed significant limitations to its widespread use. Nevertheless, the discovery of graphene has spurred research into similar new 2D materials. Silicene and germanene are two of these emerging materials. Early theoretical studies predicted that their structures would show similar electronic properties to graphene, such as a Dirac cone at the high symmetry k-point of the Brillouin zone, large charge carrier mobility, and the quantum spin Hall effect [3-5]. However, both silicene and germanene differ from graphene in that they have better opto-magnetic properties [6-8].

The major challenge has been to grow these new 2D materials, since the theoretical models indicated that freestanding germanene and silicene layers had buckled structures that were unstable compared to graphene. So far, single 2D layers of germanium and silicon have been grown on metal surfaces. In the case of germanene, the initial studies were performed on Ag(111) [9-12] and more recently on Al(111) [13-16], Au(111) [17-18] and Pt(111) surfaces [19], where several models for the structure of the germanene layer have been proposed. Following early theoretical calculations proposing the existence of silicene [4-5], the growth of silicene was investigated on Ag(111), where a 2D sheet of silicene was observed by STM measurements for the first time [20]. Subsequent studies have shown that several silicene structures can be grown on the Ag(111) surface, and they can co-exist under certain growth conditions [21-30]. At the same time, experiments were performed on the Ag(110) surface [31-35]. On the Ag(110) substrate, self-assembled silicene nano-ribbons (NRs) 1.6 nm wide were obtained [34], which were found to be weakly reactive towards oxidation [33]. Silicene has also been grown on several other metallic surfaces including Ir(111) [36], $ZrB_2(0001)$ [37], Pt(111) [38], and recently on Au(110) [39] and Au(111) [40-41]. A more in-depth description of these studies is given in two recent review articles [42-43].

In this paper we report on the complexity of the first stages of silicene growth on Ag(111). We show that there is a strong interaction with the substrate and that growth proceeds via the expulsion of Ag atoms from the top layer of the terraces. In the STM images, atomic structures are observed that are consistent with the formation of a mixed Si-Ag top layer.

2. Experiment

he STM experiments were performed at room temperature (RT) in an UHV chamber equipped with low energy electron diffraction (LEED), and a commercial Omicron STM. The sample was prepared in an ultrahigh-vacuum system with a base pressure of ~ $1x10^{-10}$ mbar. After repeated cycles of Ar^+ sputtering (600 eV and P = 5 x 10^{-5} mbar) and annealing at 450°C for a few minutes, the Ag(111) surface showed a sharp (1x1) LEED pattern. Silicon was evaporated under UHV conditions by direct current heating of a 12x5 mm Si wafer at about 1250°C. The temperature of the substrate can be controlled during the silicon deposition.

3. Results and Discussions

Figure 1a displays a 60x60 nm^2 STM topography recorded after a sub-monolayer coverage of silicon on the Ag(111) surface, held at 250°C during the deposition. The STM image shows three different domains corresponding to different silicene superstructures. The first one, identified as domain A, exhibits a periodic 4x4 structure with respect to the substrate. The second domain (B) has a (2√3x2√3)R30° structure. These two well-known superstructures correspond to a silicene sheet oriented by 0° and 30° with respect to the principle lattice directions of Ag(111). In Figure 1a, a new structure composed of silicon stripes is observed (domain C) that has not been reported in the literature. Figure 1b shows a zoom of this domain. The line scan measured along the line L gives a distance between the nearest neighbor bright protrusions of about 1nm. We observe that the stripes are either isolated or assembled in pairs (Figure 1b). In addition, in domain C, we observe a small area with a 2D hexagonal arrangement and a 1nm unit cell (Figure 1c). This latter structure corresponds to the (√13x√13)R13.9° type I already reported [22,43]. Therefore, we can conclude that these stripes are precursors during the growth of the (√13x√13)R13.9°-type I structure.

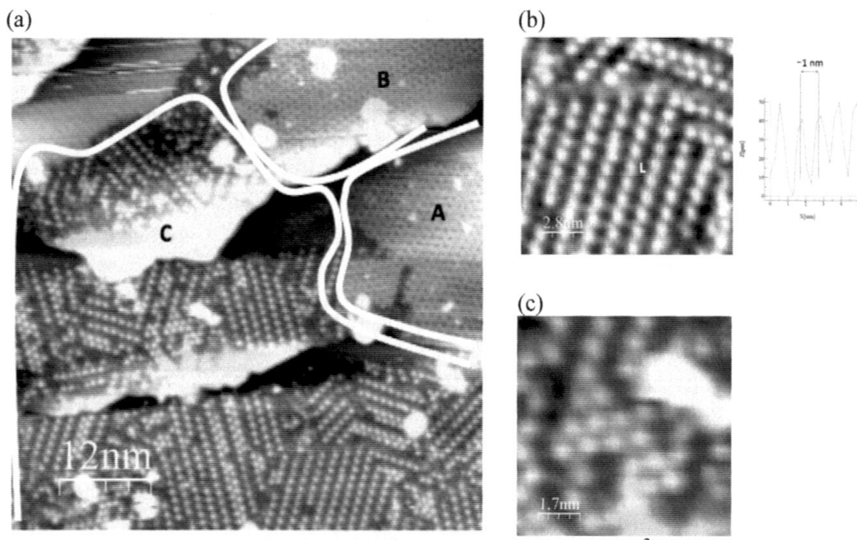

Figure 1: STM images recorded at -1.7V and 0.3 nA: (a) 60x60nm^2 image showing different domains of silicene (A, B, C), (b) and (c) zooms on domain C showing the new stripe structure, and the precursor to the (√13x√13)R13.9° type I structure, respectively.

Figure 2a shows another area of the sample. In addition to the 4x4 reconstruction and the new silicon stripes, domains corresponding to the (√13x√13)R13.9°-type II structure can be seen. It has been shown that both the type I and type II structures can coexist on the same surface [22]. However, an important new aspect is present. We observe areas of bare silver and silicene structures coexisting on the same terrace. The line scan L1 in Figure 2b shows a weak corrugation (around 0.02 nm) indicating that the silicon and Ag atoms are in the same plane. In comparison, the line scan L2 shows a step with a height of 0.22 nm height which is the expected value for an inter-atomic layer on Ag(111).

These experimental results indicate that the silicene does not form on the Ag (111) surface by adsorbing on the unmodified silver substrate. We observe that the bare silver regions are always located within the terrace, away from the step edge. In other words, below the edge of a higher terrace. In addition, some islands are observed close to the step edges or among the silicene region, but never on the bare silver surface. Therefore, we can assign these islands to the silver atoms expelled from their initial location. This can be explained by the diffusion of the expelled silver atoms across the terrace to the step edge. This implies that an exchange between the silicon and silver atoms has taken place so that the silicene structures develop on the same level as the silver. Such a scenario has been proposed in the literature [24]. It was argued that the released adsorption energy of a Si atom allows the diffusion of this atom on the Ag(111) surface and its substitution with a Ag atom of the topmost layer. The excess energy may also favor the diffusion of the expelled Ag atoms, explaining the propagation of the silver step edges during the growth of silicene [25]. Recent studies of germanium adsorption on Au(111) and Al(111) using ion scattering techniques show that the Ge atoms coexist with the metal atoms in the topmost layer [44-45]. These results support a number of previous studies that had proposed or raised the question of the formation of a 2D surface silicide or germanicide on the Ag, Al, Au, and Pt surfaces [11,16,18-19,38,40].

The present STM measurements confirm a mechanism of exchange between Si and Ag atoms leading to the growth of silicene on Ag(111) substrate at temperatures higher than room temperature. In addition, our data suggest that the first steps of silicene growth start at the step edges. The silicene domain then extends further across the whole terrace. Therefore, silicon atoms are likely to exchange with silver atoms located at the step edges. Thus, it is quite reasonable to propose that at elevated sample temperatures, expulsion and diffusion processes produce an intermediate mixed top layer that evolves into the commonly observed 2D silicene reconstructions.

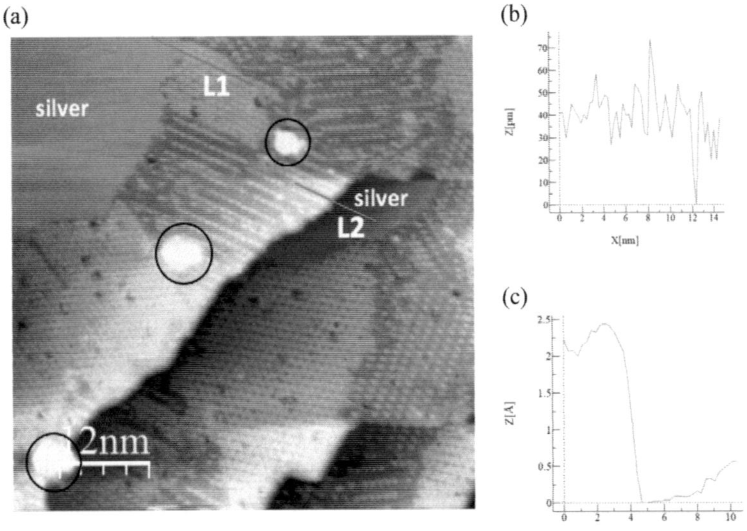

Figure 2: (a) 60x60nm^2 STM image recorded at -1.7V and 0.3 nA showing the coexistence of the bare silver with the striped silicon structures on the same terrace. The expelled silver atoms form islands that are highlighted by the black circles. (b) and (c) show the line profiles of the bare silver surface and the step height, respectively.

4. Conclusion

We have studied the growth of silicene on Ag(111) substrate in the sub-monolayer range using STM. It was possible to identify different domains of silicene, as well as domains of bare silver substrate, within the same topography. A detailed analysis of these domains reveals that silicon atoms exchange with surface silver atoms. In addition, the growth mechanism of silicene starts from the step edges of the substrate. Different nucleation sites on the step edges are probably responsible for the formation of different domains of silicene of nanometric size. This study shows that the initial growth of the silicene structures proceeds via the expulsion and diffusion of the metal atoms during an intermediate stage where a mixed topmost layer containing both Si and Ag exist. This supports the observations in a number of recent studies of mixed Si-metal or Ge-metal surface 2D alloy phases.

5. References

[1] Fasolino A, Los J H, and Katsnelson M I, 2007, *Nature Materials* **6**, 858-61
[2] Katsnelson M I, 2007 *Materials Today* **10**, 20–27
[3] Fagan S B, et al., 2000 *Phys. Rev. B* **61**, 9994
[4] Lebègue S, and Erikksson O, 2009, *Phys. Rev. B* **79**, 115409
[5] Cahangirov S, Topsakal M, Aktuerk E, Sahin H, and Ciraci S, 2009 *Phys. Rev. Lett.* **102**, 236804
[6] Rachel S, Ezawa M, 2014 *Phys. Rev. B* **89**, 195303.
[7] Matthes M, Pulci O, Bechstedt F, 2014 *New J. Phys.* **16**, 105007.
[8] Tabert C J, Nicol E J, 2013 *Phys. Rev. Lett.* **110**, 197402.
[9] Oughaddou H, Aufray B, Bibérian J P, HoarauY Y, 1999 *Surf. Sci.* **429**, 320
[10] Oughaddou H, Gay J M, Aufray B, Lapena L, Le Lay G, Bunk O, Falkenberg G, Zeysing J H, Johnson R L, 2000 *Phys. Rev. B* **61**, 5692
[11] Oughaddou H, Sawaya S, Goniakowski J, Aufray B, Le Lay G, Gay J M, Tréglia G, Bibérian J P, Barrett N, Guillot C, Mayne A J, Dujardin G, 2000 *Phys. Rev. B* **62**, 16653
[12] Oughaddou H, Mayne A, Aufray B, Bibérian J P, Le Lay G, Ealet B, Dujardin G, Kara A, 2007 *J. Nanoscience and Nanotechnology* **7**, 3189
[13] Derivaz M, Dentel D, Stephan R, Hanf M, Mehdaoui A, Sonnet P, Pirri C, 2015 *Nano Lett.* **15**, 2510–2516
[14] Stephan R, Hanf M C, Derivaz M, Dentel D, Asensio M C, Avila J, Mehdaoui A, Sonnet P, Pirri C, 2016 *J. Phys. Chem. C* **120**, 1580-1584.
[15] Fukaya Y, Matsuda I, Feng B, Mochizuki I, Hyodo T, Shamoto S, 2016 2D Mater. **3**, 035019
[16] Wang W, Uhrberg R I G, 2017 *Beilstein J. Nanotech.* **8**, 1946-1951
[17] Dávila M E, Xian L, Cahangirov S, Rubio A, Le Lay G, 2014 *New J Phys.* **16**, 095002
[18] Wang W, Uhrberg R I G, 2017 *Phys. Rev. Mat.* **1**, 074002
[19] Li L, *et al.* Buckled Germanene Formation on Pt(111). 2014 *Adv. Mater.* **26**, 4820–4824.
[20] Lalmi B, Oughaddou H, Enriquez H, Kara A, Vizzini S, Ealet B, and Aufray B, 2010 *Appl. Phys. Lett.* **97**, 223109
[21] Majzik Z, Tchalala M R, Svec M, Hapala P, Enriquez H, Kara A, Mayne A J, Dujardin G, Jelinek P, and Oughaddou H, 2013 *J. Phys.: Condens. Matter,* **25**, 225301
[22] Tchalala M R, Enriquez H, Yildirim H, Kara A, Mayne A J, Dujardin G, Ait Ali M, and Oughaddou H, 2014 *Appl. Surf. Sci.* **303**, 61–66
[23] H. Enriquez, A. Kara, A.J. Mayne, G. Dujardin, H. Jamgotchian, B. Aufray, H. Oughaddou, J. Phys. Conf. Ser. **491**, 012004 (2014)
[24] Bernard R, Borensztein Y, Cruguel H, Lazzeri M, and Prévot G, 2015 *Phys. Rev. B* **92**, 045415
[25] Satta M, Colonna S, Flammini R, Cricenti A, and Ronci F, 2015 *Phys. Rev. Lett.* **115**, 026102
[26] Meng L, Wang Y, Zhang L, Du S, Wu R, Li L, Zhang Y, Li G, Zhou H, Hofer W A, and Gao H-J, 2013 *Nano Lett.* **13**, 685.
[27] Lin C L, Arafune R, Kawahara K, Tsukahara N, Minamitani E, Kim Y, Takagi N, and Kawai M, Applied Physics Express **5**, 045802 (2012)
[28] Feng B, Ding Z, Meng S, Yao Y, He X, Cheng P, Chen L, Wu K, 2012 *Nano letters* **12**, 3507-3511
[29] Enriquez H, Vizzini S, Kara A, Lalmi B, and Oughaddou H, 2012 *J. Phys.: Condens. Matter*, **24**, 314211
[30] Jamgotchian H, Ealet B, Colignon Y, Maradj H, Hoarau J-Y, Biberian J P, Aufray B, 2016 *J. Phys.: Condens. Matter* **27**, 395002
[31] Le Lay G, Aufray B, Léandri C, Oughaddou H, Bibérian J P, De Padova P, Davila M E, Ealet B, Kara A, 2009 *Appl. Surf. Sci.* **256**, 524
[32] Kara A, Léandri C, Davila M E, De Padova P, Ealet B, Oughaddou H, Aufray B, and Le Lay G, 2009 *J. Supercond. Nov. Magn.* **22**, 259
[33] De Padova P, Léandri C, Vizzini S, Quaresima C, Perfetti P, Olivieri B, Oughaddou H, Aufray B, and Le Lay G, 2008 *Nano Lett.* **8**, 2299

[34] Aufray B, Kara A, Vizzini S, Oughaddou H, Léandri C, Ealet B, and Le Lay G, 2010 *Appl. Phys. Lett.* **96**, 183102
[35] M.R. Tchalala, H. Enriquez, A. J. Mayne, A. Kara, G. Dujardin, M. Ait Ali, H. Oughaddou, J Phys. Conf. Ser. **491**, 012002 (2014)
[36] Meng L, Wang Y, Zhang L, Du S, Wu R, Li L, Zhang Y, Li V, Zhou H, Hofer W.A, Gao H.-J, 2013 *Nano Lett.* **13**, 685-690.
[37] Fleurence A, Friedlein R, Ozaki T, Kawai H, Wang Y, Yamada-Takamura Y, 2012 *Phys. Rev. Lett.* **108**, 245501.
[38] Švec M, Hapala P, Ondráček M, Merino P, Blanco-Rey M, Mutombo P, Vondráček M, Polyak Y, Cháb V, Martín Gago J A, and Jelínek P, 2014 *Phys. Rev*. B **89**, 201412.
[39] Tchalala M R, Enriquez H, Mayne A J, Kara A, Roth S, Silly M.G, Bendounan A, Sirotti F, Greber Th, Aufray B, Dujardin G, Ait Ali M, and Oughaddou H, 2013 *Appl. Phys. Lett.* **102**, 083107.
[40] H. Enriquez, A.J. Mayne, A. Kara, S. Vizzini, S. Roth, B. Lalmi, A.P. Seitsonen, B. Aufray, Th. Greber, R. Belkhou, G. Dujardin, H. Oughaddou, 2012 Appl. Phys. Lett. **101**, 021605
[41] Sadeddine S, Enriquez H, Bendounan A, Das P K, Vobornik I, Kara A, Mayne A J, Sirotti F, Dujardin G, and Oughaddou H, 2017 *Scientific Reports,* **7**, 44400
[42] Kara A, Enriquez H, Seitsonen A P, Lew Yan Voon L C, Vizzini S, Aufray B, and Oughaddou H, 2012 *Surface Science Reports*, **67**, 1-18
[43] Oughaddou H, Enriquez H, Tchalala M R, Yildirim Y, Mayne A J, Bendounan A, Dujardin G, Ait Ali M, and Kara K 2015 *Progress in Surface Science* **90**, 46–83
[44] Endo S, Kubo O, Nakashima N, Iwaguma S, Yamamoto R, Kamakura Y, Tabata H, Katayama M, 2018 *Appl. Phys. Exp.* **11**, 015502
[45] Cantero E D, Solis L M, Tong Y, Fuhr J D, Martiarena M L, Grizzi O, Sánchez E A, 2017 *Phys. Chem. Chem. Phys.* **19**, 18580

IOP Conf. Series: Journal of Physics: Conf. Series **1081** (2018) 012006 doi:10.1088/1742-6596/1081/1/012006

Graphene oxide/Fe_3O_4 nanorods composite: structural and Raman investigation

R Hatel[1,4], S El Majdoub[1], A Bakour[1], M Khenfouch[2,3], M Baitoul[1]
[1]Group of Polymers and nanomaterials, Laboratory of Solid state Physics, Faculty of Sciences Dhar el Mahraz, University Sidi Mohammed ben Abdellah, PO Box 1796, 30 000 Atlas Fez, Morocco
[2] UNISA Africa Chair in Nanosciences-Nanotechnology, College of Graduate Studies, University of South Africa, Muckleneuk Ridge, Pretoria, South Africa
[3] iThemba LABS-National Research Foundation of South Africa, Old Faure Road, Somerset West, Western Cape Province, South Africa

E-mail : hatel.rhizlane@gmail.com

Abstract. The emerging field of organic/inorganic nanocomposites has become the focus of considerable research in recent years. In particular, graphene coated with iron oxide nanostructures are widely required for various applications. In this work, we report the preparation of graphene oxide/magnetite GO/Fe_3O_4 nanocomposite via a simple hydrothermal method. Our experimental results obtained by transmission electron microscopy (TEM) reveal that well defined Fe_3O_4 nanorods (NRs), about 8-11 nm in diameter and 50-60 nm in length, were uniformly anchored on graphene oxide nanosheets. X-ray diffraction (XRD) confirms the successful assembly of Fe_3O_4 on to GO nanosheets. Thermo gravimetric analysis results(TGA) show a good thermal stability for the nanocomposite. Raman laser studies carried out at various laser powers and acquisition times show that all the samples undergo changes in intensities and peak positions which reflect the modification of the electronic structure and electron-phonon interactions.

1. Introduction

The scientific interest in graphene, a new crystalline form of carbon consisting of a two dimensional sp^2 bonded sheet, has generated thousands of publications and an accelerating pace of research owing to its exceptional physical properties, chemical tunability, and potential for various applications [1]. Graphene can generally be produced by several disruptive technologies and each strategy has its own advantages and disadvantages; nevertheless, the production of large-area and free-standing graphene-based hybrid with high flexibility and multifunctional features is still a challenge [2]. On the other hand, graphene oxide, an atomically thin and semi-aromatic network of sp^2/sp^3 bonded carbon atoms intermittently decorated with oxygen-containing functionalities on its basal planes and in edges, is currently the ideal precursor for cost-effective and scalable production of graphene via a simple chemical method [3]. The polar lamellar surface of GO allows developing various graphene-based hybrid materials via easy chemical modification by the incorporation of various types of nanoparticles such as metal [4,5], metal oxide [6], and quantum dots [7]. Among those hybrid materials, the ones

[4] To whom any correspondence should be addressed

with the magnetic nanoparticles constitute an important class of materials due to their numerous applications in different technological field as energy storage, enhanced optical limiting, magnetic resonance imaging (MRI), drug delivery and environmental remediation [8].Our previous work reports the synthesis of grapheme derivatives decorated with Fe_3O_4, one of the most attractive supraparamagnetic materials due to its remarkable properties [9]. As results, we reveal a charge transfer and donor-acceptor interaction between the Fe_3O_4 nanostructures and the graphene derivatives which can be supported by the fact that graphene oxide is an excellent electron acceptor. Recently, significant importance has been given to develop these phenomena in graphene-based nanocomposites trough functionalization and chemical doping [10].
In the present investigation, our goal is to prepare graphene coated with Fe_3O_4 nanoparticles and to further elucidate the effect of the laser power beam on our elaborated samples using Raman spectroscopy with an excitation wavelength of 785 nm and different parameters (power and acquisition time). The structural and thermal properties are provided by TEM, XRD and TGA.

2. Experimental details

2.1. Synthesis of Graphene oxide

GO nanosheets are synthesized from natural graphite powders using a modified Hummers method as shown in Figure1. Briefly, 2 g graphite powder was added to 46 ml concentrated sulphuric acid at 0 °C, and then 6g of potassium permanganate was added gradually under magnetic stirring. The mixture was then stirred at 35 °C for 40 min, followed by slowly increase in temperature until 98 °C. Simultaneously, 92 ml of distilled water was slowly added and the mixture was maintained at that temperature for 20 min under reflux condition. At the end of the reaction, 280 ml of distilled water followed by 20 ml of 30% H_2O_2 solution was added. The solid product was separated by centrifugation, washed repeatedly with 5% HCl solution and then washed several times with acetone and water. After this, the sample was dried in a vacuum oven at 60 °C overnight [11].

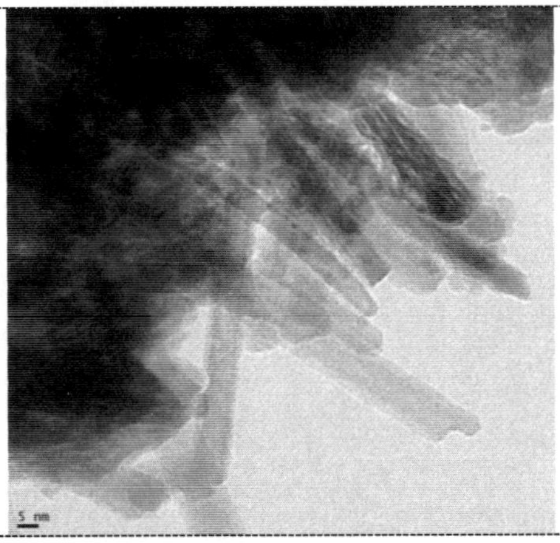

Figure 1: TEM image of GO/Fe_3O_4 nanocomposite

2.2. Synthesis of GO/Fe_3O_4 nanocomposite

Graphene oxide/Fe_3O_4 nanocomposites were synthesized through a facile chemical process as reported in our previous work [12]. For the mixture of GO with $FeSO_4.7H_2O$ and $FeCl_3$ solution, the molar ratio Fe^{3+} and Fe^{2+} was adjusted to 1.5:1. Under ultrasonic agitation, black precipitate was produced immediately by adding sodium hydroxide (NaOH). The Obtained Fe_3O_4 precipitate was stirred at 65 °C for 30 min in water bath. To purify the prepared Fe_3O_4 particles, the samples were washed repeatedly with deionized water and ethanol until pH level of 7 was reached. Particles were then dried in vacuum.

3. Results and discussion

3.1. Transmission electron microscopy (TEM)

As shown in Figure 1, the morphology of magnetite-graphene oxide was observed by TEM. Generally, controlling the size and the shape of these magnetic nanostructures is still a great challenge in research field. TEM image clearly reveals that the iron oxide nanostructures in the nanocomposite are well-defined as one dimensional (1D) nanorods with about 8-11nm in diameter and 50-60 nm in length. The Fe_3O_4 nanorods are uniformly dispersed on the surface of graphene oxide nanosheets. No serious aggregation of Fe_3O_4 nanorods is observed due to the enormous surface area of GO which provide numerous active sites for deposition of the nanorods, thus it can effectively hinder their agglomeration [13].

3.2. X-ray diffraction (XRD)

Structural properties of GO, Fe_3O_4 NRs and GO-Fe_3O_4 nanocomposites were investigated as shown in Figure 2. XRD pattern of GO has a strong diffraction peak at $2\theta=11.64°$ which originated from the diffraction on its (001) layer planes. The basal spacing (d_{basal}) is greatly expanded to 0.75 nm for GO compared to the interlayer distance of graphite 3.35 A as reported in the literature [14]. This broadnessis attributed to the introduction of various oxygen-containing functionalities and intercalated water molecules between graphite layers. For Fe_3O_4 nanorods, all diffraction peaks in the XRD pattern, at 2θ= 30.34°, 35.79°,43:26°, 53:84°, 56:81°, 63.01° and 75.32°corresponding to the reflections from the (220), (311), (400), (422), (511),(440) and (533), can be easily indexed to the pure cubic inverse spinel structure of magnetite, which matches well with the reported data (JCPDS: 63-3107) [15]. The peaks are intense and well defined which suggests a good degree of structural order at long range. As result of combination between the two components, GO/Fe_3O_4 spectrum presents a broad peak around 25° corresponding to the layer exfoliation and partial reduction of GO which confirm the dispersion of Fe_3O_4 onto graphene layers. This broadness overlapsthe majority of Fe_3O_4 peaks due to the low concentration of Fe_3O_4 in the composite.

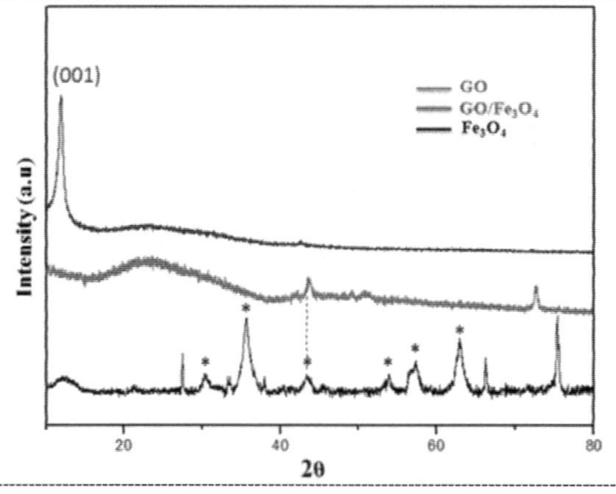

Figure 2: X-ray diffraction of GO, Fe_3O_4 and GO/Fe_3O_4nanocomposite

3.3 *Thermogravimetric analysis (TGA)*
Thermogravimetric analysis is also performed on GO, Fe_3O_4 NRs and GO-Fe_3O_4 nanocomposite to distinguish the thermal stability and confirm the attachment of Fe_3O_4 nanoparticles on the surface of graphene. Figure 3 shows TGA curve of GO, two stages appeared for mass loss with increasing temperature. The first mass loss at around 100 °C can be ascribed to the removal of the adsorbed water. The second main mass loss at around 200 °C is attributed to the decomposition of labile oxygen functional groups [16]. However, there was no significant weight loss in the TGA curve of Fe_3O_4 NRs. For the nanocomposite, there isa slow mass loss observed betweenroom temperature and 360°C. This could be attributed to the removal of the remaining water in the composite. Interestingly, the thermogravimetric profile shows a much better thermal stability for the GO/Fe_3O_4 nanocomposite [17].

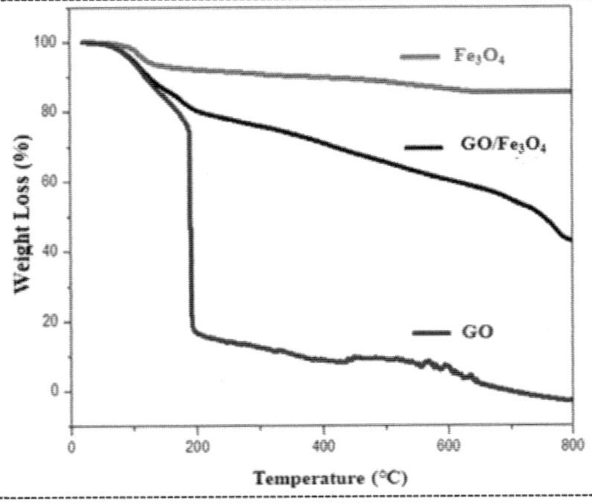

Figure 3:TG curves of GO, Fe_3O_4 and GO/Fe_3O_4

3.4. *Raman spectroscopy*

3.4.1 *Raman spectra of GO*
Raman spectroscopy affords a non-destructive technique in the study of the bonding nature of carbonaceous materials. Graphene oxide spectra exhibit two main characteristic peaks of carbon networking structures commonly known as G-band and D-band as shown in Figure 4. The G-band corresponds to the first-order scattering of the E_{2g} mode observed for sp^2 carbon domains, whereas the D-band is due to phonon branches around the K point and caused by structural effects or edges that can break the symmetry and selection rule [18]. Generally, position and relative intensity of these two bands are very sensitive to structural changes and there are several factors that can affect them, such as doping, layer numbers, exfoliation , and defects on the surface or in the edges [13]. Figure 4a depicts the Raman spectra of GO under different laser power and acquisition time. The change of these two parameters have a clear effect on the sample, by increasing the laser power D band is become larger wich can be explained by the creation of more defects as well as the two bands are red-shifted as shown in the table below, provide reliable evidence for the modification in the electronic properties of graphene due to the laser-induced electron-phonon interaction.
Moreover, It can be seen from Figure 4b that I_D/I_G increases with increasing the laser power confirming a remarkable increase in the size of the defect domains in GO.

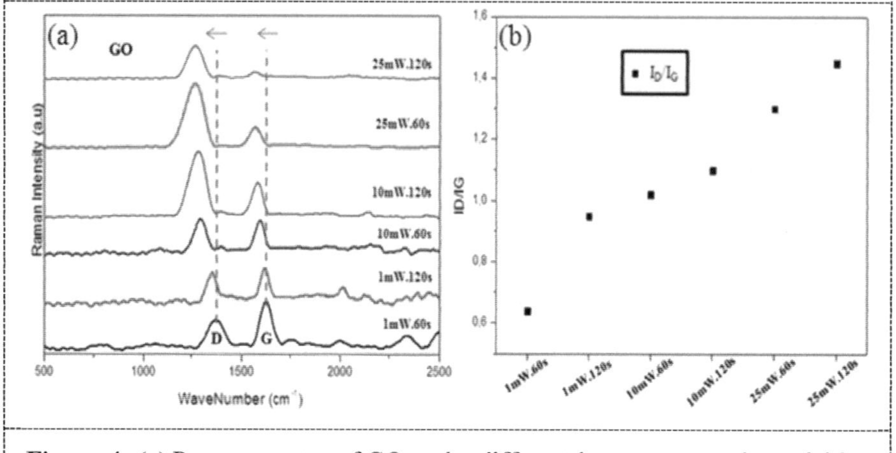

Figure 4: (a) Raman spectra of GO under different laser power and acquisition time, (b) variation of intensity ratio I_D/I

Table1: G and D position under different laser power and acquisition time

Samples(Power, time)	1mW.60s	1mW.120s	10mW.60s	10mW.120s	25mW.60s	25mW,120s
G band position	1620	1614	1587	1580	1569	1564
D band Position	1372	1356	1314	1295	1268	1265

3.4.2. Raman spectra of Fe_3O_4 nanorods

The same analysis is applied on the magnetite avoiding the high laser power (25mW) because Fe_3O_4 is a material for which special care has to be taken during the measurement of Raman spectra due to the transition metal inside its structure is divalent: ferrous/ferric (Fe^{2+}/Fe^{3+}). The high laser power focused on a small area increases the local temperature to hundreds of degrees which may causes thermal transformation, oxidation, phase transition or decomposition of the sample [19].

The theoretical analysis based on the factor-group approach predicts five Raman active bands A_{1g}, E_g and three T_{2g}, however earlier studies are reported that only four bands around 190-700 cm^{-1} are observed [20]. In Figure 5, Raman spectra of magnetite are affected by the laser power. The peaks in Raman spectrum of magnetite under 10mW and 60s are well defined and located at 290, 360, 501 and 670 cm^{-1}, respectively. The feature at 1306 cm^{-1} is assigned to a two-magnon scattering which arises from the interaction of two magnons created on anti-parallel close spin sites.

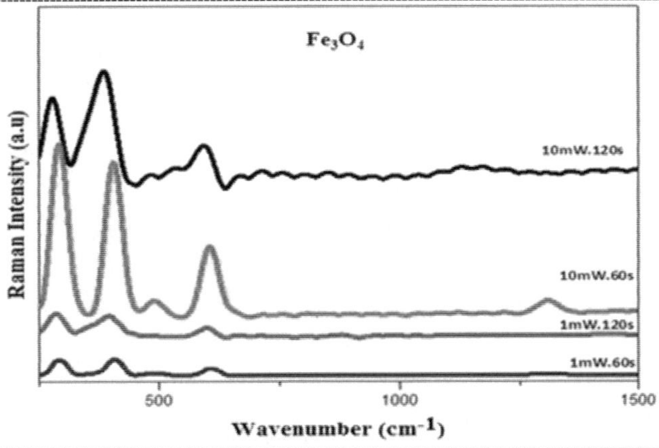

Figure 5: Raman spectra of Fe_3O_4 nanorods under different laser power and acquisition time

3.4.3. Raman spectrum of GO/Fe_3O_4 nanocomposite
As a result of the Raman study above,GO/Fe_3O_4 nanocomposite is acquired under 10 mW as power laser and 60s as acquisition time in Figure 6. Consequently, the GO/Fe_3O_4 spectrum presents the trace of the two components which confirm the functionalization of graphene oxide sheets by the magnetite nanoparticles. Obviously, magnetite peaks are in the low frequency region as well as the graphene bands are blue-shifted after incorporation with Fe_3O_4 nanoparticles. Singh K et al. attributed this blue-shifted to the charge transfer from the iron oxide to the graphene oxide [21].
After chemical functionalization of GO with Fe_3O_4 nanorods, the ratio of the intensities of the D and G bands increases from 1.10 to 1.22, which is attributed to the decrease in the mean crystallite size of graphene network by the creation of new graphitic domains that are smaller than those present in the GO. On the other hand, the FWHM of D band corresponding to the nanocomposite is less than this of GO which can be explained by the order between GO nanosheets after incorporation with Fe_3O_4 nanoparticles.

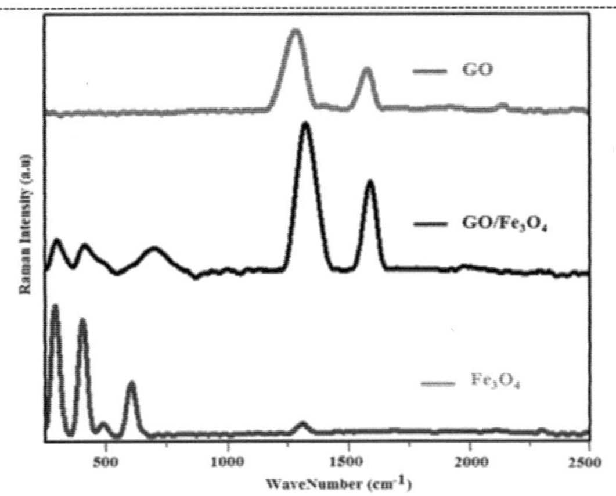

Figure 6: Raman spectrum of GO/Fe_3O_4 nanocomposite

4. Conclusion
In summary, we have successfully prepared GO/Fe_3O_4 nanocomposite via a simple chemical synthesis method. The characterization techniques were carried out using: TEM, XRD, TGA and Raman

spectroscopy. Our experimental results realized by TEM and XRD confirm the functionalization of graphene oxide nanosheets with Fe_3O_4 nanorods. The thermal stability of the nanocomposite is confirmed by TGA. Moreover, the detailed Raman studies show the influence of the laser power on the samples. The D and G bands show a sensitive response, which reveal an electron-phonon interaction caused by laser irradiation. Raman spectrum of magnetite shows four Raman activate modes in the low wave number region which confirm that Raman spectroscopy is a useful technique to identify the spinel phase of iron oxide, thus can provide us more information about interactions in graphene based materials.

References

[1] Geim A K 2009 *Science* **324** 1530-1534
[2] Allen M J, Tung V C and Kaner R B 2010 *Chem. Rev.* **110** 132-145
[3] Bielawski C W, Dreyer D R, Park S and Ruoff R S 2010 *Chem. Soc. Rev* **39** 228–240
[4] Ovid'ko I A 2014 *Rev. Adv. Mater. Sci.* **38** 190-200
[5] Khan M, Nawaz Tahir M, Adil S F, Khan H U, Siddiqui M R and Al-warthan A A2015 *J. Mater. Chem. A* **3** 18753-18808
[6] Wu Z S, Zhou G, Yin L C, Ren W, Li F and Cheng H M 2012 *Nano Energy* **1** 107-131
[7] Tian H, Shen K, Hu X, Qiao L and Zheng W 2017 *J. Alloys Compd* **691** 369-377
[8] Huang X, Yin Z, Wu S, Qi X, He Q, Zhang Q, Yan Q and Boey F 2011 *Small* **7** 1876–1902
[9] Hatel R, Khenfouch M, Baitoul M and Maaza M 2015 *Appl. Phys A* **120** 1069-1074
[10] Georgakilas V, Tiwari J N, Kemp K C, Perman J A, Bourlinos A B, Kim K S and Zboril R 2016 *Chem. Rev.* **116** 5464–5519
[11] Thema F T, Moloto M J, Dikio E D, Nyangiwe N N, Kotsedi L, Maaza M and Khenfouch M J. 2012 *Chem.* 2013 1–6.
[12] Khenfouch M, Baitoul M, Maaza M 2013 *Optical Materials* **36** 27–30
[13] Sathish M, Tomai T, Honma I 2012 *J. Power Sources* **217** 85-91
[14] Bai H, Li C, Wang X L and Shi G 2011 *J. Phys.Chem. C* **115** 5545-5551
[15] Metin Ö, Aydogan S and Meral K 2014 *J. Alloys Compd.* **585** 681–688
[16] Stankovich S, Dikin D A, Piner R D, Kohlhaas K A, Kleinhammes A, JiaY,Wu Y, Nguyen S T and Ruoff R S 2007 *Carbon* **45** 1558–1565
[17] Yang X, Zhang X, Ma Y, Huang Y, Wang Y and Chen Y 2009 *J. Mater. Chem.* **19** 2710-2714
[18] King A A, Davies B R, Noorbehesht N, Newman P, Church T L, Harris A T, Razal J M and Minett A I 2016 *Scientific Reports* **6** 19491
[19] Panta P and Bergmann C 2015 *J. Material Sci. Eng.* **5** 2169-0022. 100021
[20] Shebanova O N and Lazor P 2003 *Journal of Solid State Chemistry* **174** 424–430
[21] Singh K, Ohlan A, Pham V H, et al 2013 *Nanoscale* **5** 2411-2420

IOP Conf. Series: Journal of Physics: Conf. Series **1081** (2018) 012007 doi:10.1088/1742-6596/1081/1/012007

Empirical pseudo potential method for the energy gaps and optical properties of XN (X = Al, Ga, B) and their alloys materials *Journal of Physics: Conference Series*

L. Djoudi [a,b]* **F. Zami** [c], **M. Merabet** [a,b], **S. Benalia** [a,b], **H. Righi** [c], **D. Rached** [a]

[a]*Laboratoire des Matériaux Magnétiques, Université Djillali Liabès de Sidi Bel-Abbès, Algérie*

[b]*Centre universitaire de Tissemsilt, Institut des sciences et de la technologie, Tissemsilt, Algérie*

[c]*Laboratoire de Matière Condensée et de Développement Durable, Université Djillali Liabès de Sidi Bel-Abbès, Algérie.*

[d] *université de Batnat, Algérie*

E-mail: djoudilakhdar@yahoo.fr

Abstract. The pseudo potential method associated with the improved virtual crystal approximation (VCA) is used to calculate symmetric and anti-symmetric form factors for BN, AlN, GaN and to predict the electronic band structure of the zinc-blende $B_xAl_YGa_{1-x-y}N$ alloys lattice matched to AlN. The optical properties of these alloys are calculated. The range of compositions for which the alloys are lattice-matched to AlN substrate is determined. We find that the addition of boron to the AlGaN alloys decreases the direct gap ($E_{\Gamma\Gamma}$) and a direct to indirect transition occurs at a Al concentration of about 0.5 (y = 0.5), which corresponds to a band gap energy of about 4.75 eV.

1. Introduction

The III-nitride semiconductor, AlN, GaN, and their alloys have wide direct band gaps, which vary from 3.3 eV for GaN to 6 eV for AlN and are indispensable for the development of optoelectronic and electronic devices. Their alloys have been exploited to develop the high-power microwave transistors and the deep ultraviolet (DUV) optical devices operating at wavelength in the range of (344nm - 200nm), including light-emitting diodes and laser diodes ranging from the visible spectrum [1-4], optically pumped lasers, and photo detectors [5-8]. BN, on the other hand, has features of high thermal conductivity suitable for applications in electronics devices [9]. All these materials also have potential for high-temperature, high-power, and high-frequency electronic devices, due to their wide band gaps [10]. The (BAlGaN) quaternary system can be lattice matched to 6H–SiC and AlN substrates [11], whereas the band-gap energy (Eg) of the BAlGaN lattice matched to AlN and 6H–SiC substrates respectively is in the range of (3.6 eV–6.2 eV) and (3.8 eV– 6.2 eV) [12].

IOP Conf. Series: Journal of Physics: Conf. Series **1081** (2018) 012007 doi:10.1088/1742-6596/1081/1/012007

Some groups have studied the BAlGaN system. Takano et al. [13] have used Low-Pressure Metal-Organic Vapor-Phase Epitaxy (LP-MOVPE) (to grow (BAlGaN) and (BAlGaN/AlN) where triethylboron (TEB), trimethylaluminium (TMAl), trimethylgallium (TMGa) and ammonia (NH3) were used as the source materials for boron, aluminium, gallium and nitrogen, respectively, and used Auger electron spectroscopy (AES) and X-ray diffraction (XRD) to determine the compositions of $B_xAl_yGa_{1-x-y}N$. Here the boron composition was founded to increase from 1% to 13%. Djoudi et al [14] have used the Full potential linearized augmented plane wave (FP-LAPW) to study the structural and electronic properties of the quaternary $B_xAl_yGa_{1-x-y}N$, lattice matched to AlN substrate. Djoudi et al reported that B compositions in a $B_xAl_yGa_{1-x-y}N$ /AlN vary in a range 0 % to 14.2 % and the band gaps are, on the whole, underestimated by about 51% to 56% compared with experimental data [14, 15]. In order to understand and control the materials and device properties, we have carried out another theoretical study of the energy band gaps and optical properties of the quaternary $B_xAl_yGa_{1-x-y}N$, lattice matched to AlN substrate. We use the empirical pseudo potential method (EPM) within the virtual crystal approximation (VCA) to calculate the electronic structure of the quaternary alloy system. The empirical pseudo potential method (EPM) was developed in the 1960's [16-18] as a way to solve Schrodinger's equation for bulk crystals without knowing exactly the potential experienced by an electron in the lattice. The EPM has been one of the most popular method to calculate the full band structure for device simulation. In this study, we considered the zinc-blend poly type as a model system. Calculations are performed over the entire composition range of x and y.

The plan of the present paper is as follows: a description of the method as well as some details of the calculations in section 2. In Section 3, the energy band gaps and optical properties of quaternary $B_xAl_yGa_{1-x-y}N$ matched to AlN substrate are presented and discussed. Conclusion of the present study will be given in section 4.

2. Calculation Method

The electronic structure of the quaternary $B_xAl_yGa_{1-x-y}N$, lattice matched to AlN substrate is calculated using the empirical pseudo potential method (EPM) within the virtual crystal approximation (VCA). The fundamental concept involved in a pseudo-potential calculation is that the ion core can be omitted. Computationally this is crucial for it means that the deep core potential has been removed and a simple plane wave basis will yield rapid convergence [19]. The main advantage of using pseudopotentials is that only valence electrons have to be considered. The core electrons are treated as if they are frozen in an atomic-like configuration. As a result, the valence electrons are thought to move in a weak one-electron potential. The pseudopotential method is based on the orthogonalized plane wave (OPW) method due to Herring [20]. In this method, the crystal wave function is constructed to be orthogonal to the core states. Here the pseudo potentials is based by a set of atomic form factors Va (G) (G is a reciprocal lattice vector). The form factors are treated as adjustable parameters that can be fit to experimental data, hence the name empirical pseudopotential method. Another advantage of the empirical pseudopotential method is that only six form factors are needed to describe the band structure, where the prefactors are referred to as the symmetric ($V_S(3)$, $V_S(8)$ and $V_S(11)$) and antisymmetric ($V_A(3)$, $V_A(4)$ and $V_A(11)$) structure factors. More details about the approach has been described in Ref. [19, 21, 22]. The band gap energies, lattice constants and form factors (symmetric (V_S) and antisymetric (V_A)) for BN, AlN and GaN compounds are shown in Table 1 and Table 2, respectively. Generally, the experimental values used in the study are selected among the more recent measured band gap energies. When the experimental data are not available, we have taken instead of them the existing theoretical data in the literature.

Table 1. Band gaps of the zinc-blende structure, for BN, AlN and GaN used in the calculation.

	BN [23]	AlN[24]	GaN[25]
$E_{g\ (\Gamma-\Gamma)}$ (eV)	9.09	6	3.3
$E_{g\ (\Gamma-X)}$ (eV)	4.24	4.9	4.57
$E_{g\ (\Gamma-L)}$ (eV)	8.10	9.3	6.04

Table 2. Symmetric ($V_S(G)$) and antisymmetric ($V_A(G)$) pseudopotential form factors and lattice constants (a) used in the calculation for BN, AlN and GaN alloys.

Materials	a (a.u.)	$V_S(3)$	$V_S(8)$	$V_S(11)$	$V_A(3)$	$V_A(4)$	$V_A(11)$
BN [26]	6.8241	-0.755	0.182	0.133	0.1755	0.1250	0.0251
AlN [24]	8.26087	-0.309603	0.112783	0.067538	0.28	0.33	0.015
GaN [25]	8.50662	-0.347240	-0.016	0.21217	0.159988	0.2	0.135

3. Results and Discussions

3.1 Matching conditions

As starting point we have computed the lattice matching conditions for the $B_xAl_yGa_{1-x-y}N$ quaternary system on an AlN substrate. The lattice matching found is: x = 0.146(1-y), for which Vegard's rule has been assumed to calculate the lattice constant of the quaternary alloy under study.

$$a(x, y) = xa_{BN} + ya_{AlN} + (1\text{-}x\text{-}y)a_{GaN} \quad (1)$$

Where a(x, y), a_{BN}, a_{AlN} and a_{GaN} are the respective lattice constants of BAlGaN, BN, AlN and GaN. The composition range for which the lattice constants of $B_xAl_yGa_{1-x-y}N$ alloys are lattice matched to an AlN substrate is represented by the solid line in Figure. 1. The calculated lattice matched of $B_xAl_yGa_{1-x-y}N$/AlN alloys is in good agreement with available experimental results [13]. Here the lattice mismatch of the material under study is given by: $(a_{BAlGaN} - a_{AlN}).100\% / a_{AlN}$. In Table 3, we summarize the lattice mismatch for the $B_xAl_yGa_{1-x-y}N$ alloys for different B and Al compositions. The results clearly show that the lattice mismatch decreases when boron is incorporated with the small amounts.

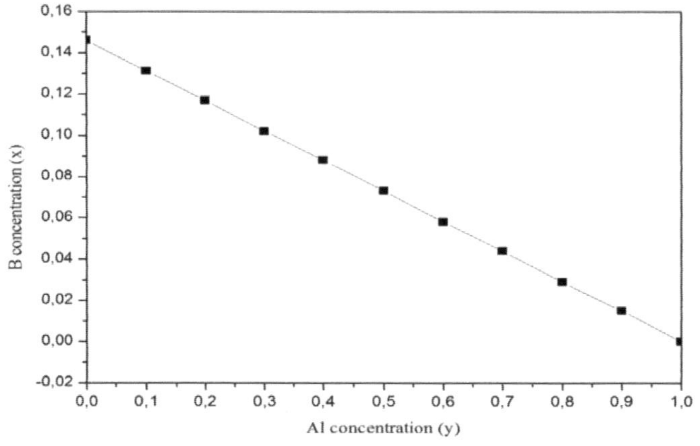

Figure 1. Solid composition of boron versus Solid composition of aluminium in the $B_xAl_yGa_{1-x-y}N$ quaternary system (14.69 % boron is the maximum composition in the BAlGaN quaternary system lattice matched to AlN).

Table 3. Lattice mismatch for the $B_xAl_yGa_{1-x-y}N$ alloys for different B and Al compositions.

Materials	**Lattice mismatch to AlN (%)**
$B_{0.3}Al_{0.06}Ga_{0.64}N$	3.314
$B_{0.03}Al_{0.06}Ga_{0.91}N$	2.185
$B_{0.03}Al_{0.10}Ga_{0.87}N$	2.066
$B_{0.06}Al_{0.20}Ga_{0.74}N$	1.158

3.2 Form factors and Band gap energies

In the next step, we have computed the form factors of the BN, AlN, GaN and $B_xAl_yGa_{1-x-y}N$ alloys lattice matched to AlN. Therefore Vegard's rule has been assumed to calculate the form factors for quaternary $B_xAl_yGa_{1-x-y}N$, so the form factors for quaternary $B_xAl_yGa_{1-x-y}N$ alloy are obtained as:

$$V_{alloy} = x\left(V_{BN}(G) - V_{GaN}(G)\right) + y\left(V_{AlN}(G) - V_{GaN}(G)\right) + V_{GaN}(G) \quad (2)$$

Where G is the reciprocal lattice vector, and V_{BN}, V_{AlN}, and V_{GaN} are the form factors for AlN, GaN and BN, respectively. Table 4 present the results found for the form factors for BN, AlN, GaN, $B_{0.03}Al_{0.06}Ga_{0.91}N$, $B_{0.03}Al_{0.10}Ga_{0.87}N$ and $B_{0.06}Al_{0.20}Ga_{0.74}N$. In Table 5, we summarize the Band gaps of the zinc-blende structure, for BN, AlN and GaN alloys found in the calculation.

Table 4. Symmetric ($V_S(G)$) and antisymmetric ($V_A(G)$) pseudopotential form factors for BN, AlN, GaN, $B_{0.03}Al_{0.06}Ga_{0.91}N$, $B_{0.03}Al_{0.10}Ga_{0.87}N$ and $B_{0.06}Al_{0.20}Ga_{0.74}N$

Materials	$V_S(3)$	$V_S(8)$	$V_S(11)$	$V_A(3)$	$V_A(4)$	$V_A(11)$
BN	-0.808610	0.294693	0.125894	-0.394826	0.125	0.0251
AlN	-0.309345	0.112795	0.067500	0.28	0.33	0.015
GaN	-0.347024	-0.016	0.212137	0.159960	0.2	0.135
$B_{0.03}Al_{0.06}Ga_{0.91}N$	-0.359	$1.048*10^{-3}$	0.201	0.151	0.206	0.125
$B_{0.03}Al_{0.10}Ga_{0.87}N$	-0.357	$6.2*10^{-3}$	0.195	0.155	0.211	0.12
$B_{0.06}Al_{0.20}Ga_{0.74}N$	-0.367	0.028	0.178	0.151	0.221	0.104

Table 5. Band gaps of the zinc-blende structure, for *BN*, *AlN* and *GaN* alloys found in the calculation.

Band gaps	BN	AlN	GaN
$E_{g\ (\Gamma\text{-}\Gamma)}$ (eV)	9.09011	6.000453	3.300499
$E_{g\ (\Gamma\text{-}X)}$ (eV)	4.24004	4.900563	4.570699
$E_{g\ (\Gamma\text{-}L)}$ (eV)	8.09994	9.300263	6.040429

A good knowledge of electronic band structure in semiconductors provides valuable information as regarding their potential utility in fabricating electronic and optoelectronic devices. In this context we have calculated the electronic band structure properties of $B_xAl_yGa_{1-x-y}N$ alloys lattice matched to AlN in the zinc-blende structure at this equilibrium lattice constants using the pseudo potential method associated with the improved virtual crystal approximation. The energy variation of the conduction band edges at Γ, X and L with respect to the top of the valence band are obtained as a function of aluminium composition. The obtained direct and indirect band-gaps are presented and plotted in figure. 2. The results shows the dependence of direct gap energy $E_{\Gamma\Gamma}$ and indirect gap energies $E_{\Gamma X}$ and $E_{\Gamma L}$ for $B_xAl_yGa_{1-x-y}N$/ AlN, on the aluminium content, where the energy band gaps $E_{\Gamma\Gamma}$ and $E_{\Gamma L}$ increase monotonically with increasing the composition (y), but ~~for~~ the indirect gap energy $E_{\Gamma X}$ varies non-monotonically with increasing the aluminium content. A direct to indirect transition occurs at an aluminium concentration of about 0.5 (y = 0.5), which corresponds to a band gap energy in the order of 4.75 eV. Above this concentration (y > 0.5), the alloy becomes an indirect gap semiconductor. The curves of figures obey the following variations:

$$E_{\Gamma-\Gamma}(eV) = 4.00227 + 1.27108y + 0.74056y^2 \quad (3)$$
$$E_{\Gamma-X}(eV) = 5.12018 - 1.03021y + 0.811667y^2 \quad (4)$$
$$E_{\Gamma-L}(eV) = 6.50422 + 2.0324y + 0.7824y^2 \quad (5)$$

These expressions may be useful for obtaining the direct and indirect band gaps for any y concentration in $B_xAl_yGa_{1-x-y}N$ alloy lattice matched to AlN. We see that the three equations are characterized by bowing parameter, for the first equation (3) the coefficient of y^2 is 0.74056 eV and for the second equation (4) is 0.811667 eV and for the last equation (5) is 0.7824 eV. This indicates that the direct gap and indirect gaps of zinc-blende $B_xAl_yGa_{1-x-y}N$/AlN has a very large deviation from the linear Vegard's law.

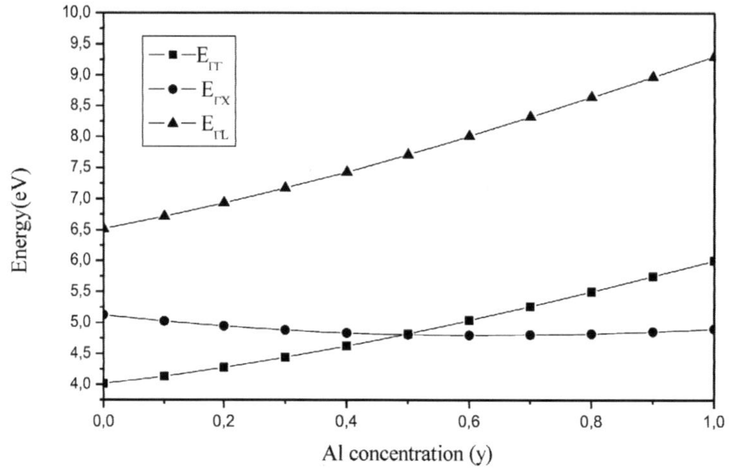

Figure 2. The energy band-gaps of $B_xAl_yGa_{1-x-y}N$ quaternary system for various aluminium concentrations

In Table 6, we have listed the fundamental band gaps (E_0), at the high-symmetry point Γ in the Brillouin zone, for the systems investigated and compared the results with available experimental data. The results clearly show that the band gaps are, on the whole, underestimated by about 11% to 17% compared with experimental data, We note that this method gives good results in comparison with other calculation method based on DFT [14, 15, 27].

Table 6. Band gaps of the zinc-blende structure, for $B_{0.03}Al_{0.06}Ga_{0.91}N$, $B_{0.03}Al_{0.10}Ga_{0.87}N$ and $B_{0.06}Al_{0.20}Ga_{0.74}N$

Material	**$E_{g\ (\Gamma-\Gamma)}$ (eV)**	
	Present calculations	**Exp. [13]**
$B_{0.03}Al_{0.06}Ga_{0.91}N$	4.081	4.92
$B_{0.03}Al_{0.10}Ga_{0.87}N$	4.137	4.73
$B_{0.06}Al_{0.20}Ga_{0.74}N$	4.286	4.84

3.3 Optical properties

In the last step, we have calculated the optical properties of $B_xAl_yGa_{1-x-y}N$. To really understand the optic properties of the semiconductors, it was necessary to know the electronic structure of these materials. The refractive index (n) for $B_xAl_yGa_{1-x-y}N$ has been calculated using the empirical expression proposed by Herve and Vandamme [28], the optical high frequency dielectric constant it is the square of a the refractive index of the material of interest and the static dielectric constant has been calculated using the empirical expression proposed by Harrison model [29]. We summarize in the

IOP Conf. Series: Journal of Physics: Conf. Series **1081** (2018) 012007 doi:10.1088/1742-6596/1081/1/012007

table 7 most of the models used to calculate the refractive index (n), the optical high frequency dielectric constant and the static dielectric constant. These models have been preferred to other models because they are found to give better agreement with known data obtained in the case of III–V semiconductors alloys [22]. The knowledge of the refractive index plays an important role in the design of heterostructure lasers, in optoelectronic devices as well as in solar cell applications. Our results (refractive index, optical high frequency dielectric constant, static dielectric constant) for various Aluminium concentrations are listed in Table 8.

Table 7. Models used to calculate the refractive index (n), the optical high frequency dielectric constant and the static dielectric constant.

Models	Observation
$n = \sqrt{1 + \left(\frac{A}{E_g + B}\right)^2}$	Eg is the fundamental energy band-gap and *A* and *B* are constants with values of 13.6 and 3.4 eV, respectively. [30,31]
$\varepsilon_\infty = n^2$	
$\frac{\varepsilon_0 - 1}{\varepsilon_\infty - 1} = 1 + v$	Harrison model. [29]
$v = \frac{\alpha_p^2(1 + 2\alpha_c^2)}{2\alpha_c^4}$	α_p is the polarity and α_c is the covalency of the material
$\alpha_p = -\frac{V_A(3)}{V_s(3)}$	α_p is the polarity proposed by Vogl [32] and $V_S(3)$ and $V_A(3)$ are the symmetric and antisymmetric pseudo potential form factors.
$\alpha_c^2 = 1 - \alpha_p^2$	

The variations of the refractive index, the high frequency dielectric constants and the static dielectric constant as a function of alloy composition are plotted in Figure. 3. It is clearly seen that the static dielectric constant decreases rapidly, the high frequency dielectric constant decreases slowly and the refractive index decreases very slowly as a function of Al concentration, respectively. The curves of figures obey the following variations:

$$n = 2.0985 - 0.33809y \quad (6)$$
$$\varepsilon_\infty = 4.38545 - 1.30255y \quad (7)$$
$$\varepsilon_0 = 7.42191 - 3.11473y \quad (8)$$

These expressions may be useful for obtaining the refractive index, the high frequency dielectric constants and the static dielectric constant for any y concentration in $B_xAl_yGa_{1-x-y}N$ alloy lattice matched to AlN. We see that the three equations are characterized by negative slope parameter, for the first equation (6) the coefficient of *y* is -0.33809 and for the second equation (7) is -1.30255 and for the last equation (8) is -3.11473. This indicates that the refractive index, the high frequency dielectric constants and the static dielectric constant of zinc-blende $B_xAl_yGa_{1-x-y}N$/AlN decreases as a function of Al concentration.

Table 8. Values of the refractive index, the static dielectric constant and the high frequency dielectric constant for different values of the Al concentration.

Aluminium (y) concentration	Refractive index	High frequency dielectric constant	Static dielectric constant
0	2.089	4.364	7.435
0.1	2.064	4.259	7.147
0.2	2.034	4.137	6.807
0.3	2.002	4.008	6.487
0.4	1.968	3.874	6.164
0.5	1.933	3.737	5.842
0.6	1.897	3.597	5.503
0.7	1.861	3.463	5.202
0.8	1.826	3.335	4.917
0.9	1.791	3.209	4.628
1.0	1.759	3.093	4.378

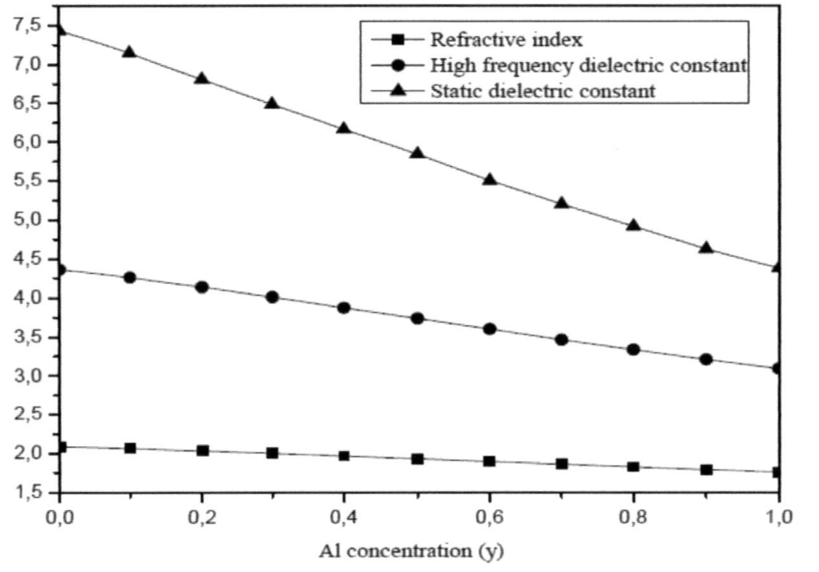

Figure 3. Variations of refractive index, the high frequency dielectric constants and the static dielectric constant as a function of alloy composition for various aluminium concentrations

4. Conclusion

The pseudo potential method (EPM) associated with the improved virtual crystal approximation (VCA) is utilized to investigate the symmetric and anti-symmetric form factors and energies band gaps for BN, AlN, GaN and $B_xAl_yGa_{1-x-y}N$ alloys lattice matched to AlN and the optical properties of this material. From the calculated results, it is found that the band gaps of the zinc-blende structure of each binary (BN, AlN and GaN) is in reasonable agreement with the experimental values, and for zinc-blende $B_xAl_yGa_{1-x-y}N$ alloys lattice matched to AlN the band gaps are, on the whole, underestimated by about 11% to 17% compared with experimental data. The absorption at the optical gaps suggested that a transition from an direct to indirect band gap may occur in $B_xAl_yGa_{1-x-y}N$ alloys lattice matched to AlN at y equal 0.5. Therefore for this material to be useful in optical applications it is necessary to keep the concentration for Aluminium less than 50%. The calculated band-gap energy of $B_xAl_yGa_{1-x-}$

IOP Conf. Series: Journal of Physics: Conf. Series **1081** (2018) 012007 doi:10.1088/1742-6596/1081/1/012007

$_y$N alloys lattice matched to AlN is in the range of 4 eV to 4.3 eV, corresponding to the wavelength range of 310 nm to 288 nm. Therefore, the BAlGaN system is a promising material for use in semiconductor lasers that operate in the ultraviolet (UV) spectral region. Our findings indicate that the refractive index, the high frequency dielectric constants and the static dielectric constant of zinc-blende $B_xAl_yGa_{1-x-y}N/AlN$ decreases as a function of Al concentration. Finally our results are predictions and may serve as reference for future experimental work.

References

[1] S. Nakamura, T. Mukai, M. Senoh, Jpn. J. Appl. Phys. 30 (1991) L1998.
[2] S. Nakamura, Science 281 (1998) 956.
[3] S. Nakamura, M. Senoh, S. Nagahama, N. Iwasa, T. Yamada, T.Matsushita, Y. Sugimoto, H. Kiyoku, Appl. Phys. Lett. 69 (1996) 4056.
[4] S. Nakamura, M. Senoh, S. Nagahama, N. Iwasa, T. Yamada, T. Matsushita, H. Kiyoku, Y. Sugimoto, Jpn. J. Appl. Phys. 35(1996) L74.
[5] Y. Taniyasu, M. Kasu, T. Makimoto, Nature 441 (2006) 325.
[6] V. Adivarahan, W.H. Sun, A. Chitnis, M. Shatalov, S. Wu, H.P. Maruska, M.A. Khan, Appl. Phys. Lett. 85 (2004) 2175.
[7] T. Detchprohm, X.Li, S.-C. Shen, P.D. Yoder, R.D. Dupuis semiconductors and Semimetals, Volume 96 (2017)
[8] T. Takano, M. Kurimoto, J. Yamamoto, H. Kawanishi, J. Cryst. Growth. 237-239 (2002) 972-977.
[9] T. Takayama, M. Yuri, K. Itoh, T. Baba, J.S. Harris Jr, J. Cryst. Growth. 222 (2001) 29-37.
[10] S. Nakamura, The Blue Laser Diode-GaN Based Light Emitters and Lasers (Springer,Berlin, 1977).
[11] M. Haruyama, T. Shirai, H. Kawanishi, Y. Suematsu, in: Proc. Int. Symp. on Blue Laser and Light Emitting Diodes, Chiba University, Chiba 1996, p. 106.
[12] M. Tsubamoto, T. Honda, J. Yamamoto, M. Kurimoto, M. Shibata, M. Haruyama, H. Kawanishi, in: Proc. Second Int. Conf. on Nitride Semiconductors,Tokushima 1997, p. 250.
[13] T. Takano, M. Kurimoto, J. Yamamoto, H. Kawanishi, J. Cryst. Growth. 237-239, 972 (2002).
[14] L. Djoudi, A. Lachebi, B. Merabet, H. Abid, Modern Physics Letters B, Vol. 26, No. 24 (2012) 1250159.
[15] L. Djoudi, A. Lachebi, B. Merabet, H. Abid, Acta Physica Polonica A, No. 4, Vol. 122 (2012).
[16] J.C. Phillips, "Energy-Band Interpolation Scheme Based on a Pseudopotential," Phys. Rev. 112, 685 (1958).
[17] J.C. Phillips and L. Kleinman, "New Method for Calculating Wave Functions in Crystals and Molecules," Phys. Rev. 116, 287 (1959).
[18] L. Kleinman and J.C. Phillips, "Crystal Potential and Energy Bands of Semiconductors. III. Self-Consistent Calculations for Silicon," Phys. Rev. 118, 1153 (1960).
[19] J.R. Chelikowsky and M.L. Cohen, Phys. Rev. B., 14, 556 (1976).
[20] Herring, C. "A New Method for Calculating Wave-Functions in Crystals" Phys. Rev. 57. 1169. 1940.
[21] N. Bouarissa. Physica B 399 (2007) 126–131.
[22] A. Hafaiedh, N. Bouarissa. Materials Chemistry and Physics 115 (2009) 122–125.
[23] F. Chiker, B. Abbar, B. Bouhafs and P. Ruterana, Phys. Stat. Sol. (b), 241, (2004), 305.
[24] A. Bhouri, F. Ben Zid, H. Mejri,A. Ben Fredj, N. Bouarissa, M. Said, J. Phys. Condens. Matter 14 (2002) 7017.
[25] N. Bouarissa, K. Kassali, Superlattice Microstruct. 35 (2004) 115.
[26] Tsay Y.F., Vaidyanathan A., Mitra M.S., Phys. Rev. B, 1979, 19, 5422.

[27] L. Djoudi, M. Merabet, F. Dahmane, M. Boucharef, S. Benalia, D. Rached, Journal of Physics: Conference Series 758 (2016) 012025.
[28] P.J.L. Herve, L.K.J. Vandamme, Infrared Phys. Technol. 35 (1994) 609.
[29] S.Y. Davydov, S.K. Tichonov, Semiconductors 32 (1998) 947.
[30] V.P. Gupta, N.M. Ravindra, Phys. Status Solidi B 100 (1980) 715.
[31] N.M. Ravindra, S. Auluck, V.K. Srivastava, Phys. Status Solidi B 93 (1979) k155.
[32] P. Vogl, J. Phys. C 11 (1978) 251.

Half-Metallic Ferrimagnetism in the Ti_2YAl (Y=Fe, Ni) Heusler Alloys

Mohamed Drief[1,2], Youcef Guermit[1], Nour-Eddine Benkhettou[1] and Djamel Rached[1]

[1] Laboratoire des Matériaux Magnétiques, Faculté des Sciences, Université Djillali Liabés de Sidi Bel-Abbés 22000, Algeria

E-mail: driefed80@gmail.com

Abstract. Using density functional theory with the full-potential linearized augmented plane wave method, we study the structural, electronic and magnetic properties of Ti_2FeAl and Ti_2NiAl alloys with Hg_2CuTi-type structure. These alloys are found to be half-metallic ferrimagnets. In addition, the total magnetic moments of Heusler alloys are Ti_2YAl (Y = Fe, Ni) agree with the Slater-Pauling rule (M_{tot} = Z_{tot} - 18) with respective gap energies of 0.56 eV and 0.44 eV in the spin down channel.

1. Introduction

Given their possible applications in spintronic devices, such as Magnetic Random Access Memory (MRAM) [1-3], half-metals (HM) are materials having a complete spin polarization (100 %) at the Fermi level, attracting then a great interest in the research field. The half-metal properties have been observed in many materials, such as Heusler compounds [4-9], metal oxides [10-12], dilute magnetic semiconductors [13, 14] and zinc-blende pnictides and chalocogenides of transition metal compounds [15-19]. Intermetallic ternary Heusler alloys are of the X_2YZ form, where X, Y are transition metals and Z is an element of the main group [20]. The X_2YZ Heusler compounds crystallize in the cubic $AlCu_2Mn$-type structure with the space group Fm3m. In this structure, X, Y and Z atoms are respectively placed on the Wyckoff positions 8c (1/4,1/4,1/4), 4a (0,0,0) and 4b(1/2,1/2,1/2).
If the number of 3d electrons in the atom Y is greater than X, $CuHg_2Ti$-type structure with the 216 F-43m space group will be observed. In this structure, the X atoms occupy the 4a (0, 0, 0) and 4c (1/4, 1/4, 1/4) positions, while Y and Z are located respectively in 4b (1/2,1/2,1/2) and 4d (3/4,3/4,3/4) sites. Many Heusler compounds have been theoretically predicted to be half-metals [5-10, 22-26].
Since calculations of electronic structure compounds based on Heusler Ti_2 have not been widely studied until now, our work presents a systematic study of the electronic and magnetic properties of a structure Ti_2YAl series (Y=Fe, Ni) alloys with Hg_2CuTi-type structure using the Full-Potential Linearized Augmented Plane Wave (FP-LAPW) method [30].
In following, we begin by explaining the methods and the calculation models, and then we present and discuss the obtained results on the structural, magnetic and electronic properties.

[2] To whom any correspondence should be addressed.

2. Computational Method

The electronic structure calculations were performed using the FP-LAPW [31] approach within the density functional theory (DFT) [32, 33] implemented in the WIEN2k code [34]. The exchange-correlation potential was treated with the generalized gradient approximation of Wu and Cohen (WC-GGA) [35] which give good results that are consistent with earlier calculations. In the FP-LAPW method, the real space is divided into two regions: non-overlapping spheres of Muffin-Tin radius RMT around the atomic sites (region1) and the interstitial region (region 2). In region 1, wave functions, electron density and potential are expanded in combination of spherical harmonics up to l_{max} =10, in region 2 in Fourier series, or plane waves, with a cutoff radius RMT K max= 8. The values of RMT used in the present work are listed in Table 1. The magnitude of the largest vector responsible in the Fourier expansion of density (G_{max}) was 14 $(a.u.)^{-1}$. A mesh of 84 K-point was made in the irreducible wedge of the Brillouin zone. The cutoff energy, which defines the separation of valence and core states, was chosen as -81.63 eV. The self-consistent calculations were considered to have converged when the total energy was stable to within 1.36 meV.

Table 1. RMT of the Ti_2FeAl and Ti_2NiAl alloys.

Alloy	Ti(1)	Ti(2)	Fe	Ni	Al
Ti_2FeAl	2.28	2.28	2.40	-	2.17
Ti_2NiAl	2.33	2.33	-	2.45	2.21

3. Results and discussion

The Heusler alloys produced with the formula X_2YZ crystallize either in $AlCu_2Mn$-type or $CuHg_2Ti$, typical for the full-Heusler alloys with Wyckoff position of the atom X occupying A (0,0,0) and B (1/4,1/4,1/4) sites. The Y atom occupies C (1/2,1/2,1/2) site and the atom Z D (3/4,3/4,3/4) site. In our Ti_2YAl (Y=Fe, Ni) compound, the atomic number of Fe is greater than that of Ti. The ground state properties of Ti_2YAl alloys with $CuHg_2Ti$ structure with ferrimagnetic and ferromagnetic type are obtained by calculating the total energy of these systems as a function of the volume of the unit cell using the state Murnaghan equation of state [36] to fit the energies. The results from the calculations are shown in Table 2.

Table 2. Calculated and literature lattice constants a, bulk modulus B, first derivatives B' and gap energies ΔE for Ti_2FeAl and Ti_2NiAl compounds.

Alloy	Type	a (Å) Calculated	a (Å) Literature	B (GPa) Calculated	B (GPa) Literature	B' Calculated	B' Literature	ΔE (meV)
Ti_2FeAl	Ferro	6.059	6.069 [26]	151.15	134.138 [26]	3.72	6.0915 [26]	- 0.48
	Ferri	6.062	5.99 [27]	142.69	160.539 [27]	4.66	4.675 [27]	
Ti_2NiAl	Ferro	6.2051	6.195 [30]	124.337		3.995		- 0.68
	Ferri	6.0226	6.20 [28,29]	124.923		3.923		

One can see that the volume of the ferrimagnetic state is more stable than the ferromagnetic state, as shown in Figure 1. The energy difference is -0.48 meV for Ti_2FeAl and -0.68 meV for the Ti_2FeAl.

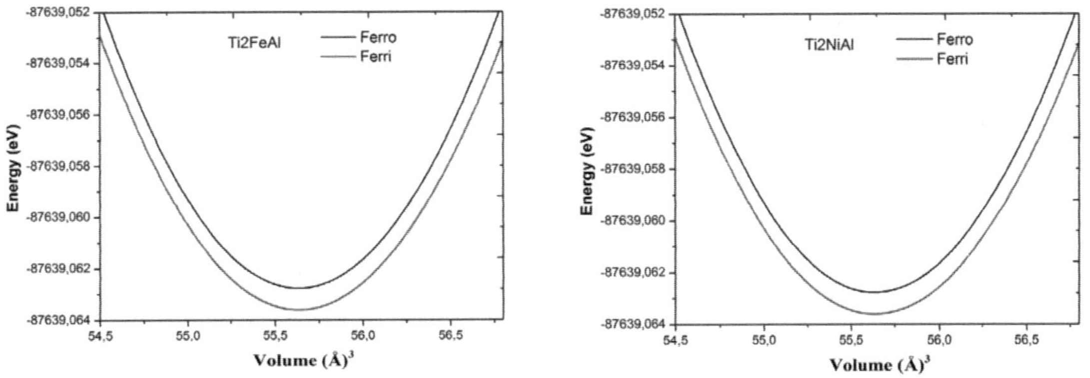

(a) (b)

Figure 1. Total energy of Ti2YAl as a function of volume the ferromagnetic and ferrimagnetic phase for (a) the Ti_2FeAl compound, (b) the Ti_2NiAl compound.

In figure 2, we present the calculated spin-resolved band structures of Ti_2YAl (Y=Fe, Ni) alloys. Since the Fermi level is pinned at majority spin states, the band structure has a metallic character, whereas the Fermi level is located in a band gap in the minority spin state: the band structures have then a semiconducting characteristic. Therefore, a spin polarization of 100% is located at the Fermi level of Ti_2YAl (Y=Fe, Ni) alloys and a complete half-metal (HM) for the equilibrium constant of the lattice from the partial density of states (PDOS) for Ti_2YAl (Y=Fe, Ni) alloys is shown in figure 3. One can see that the energy bands in Ti_2YAl (Y=Fe, Ni) alloys are divided into three parts in both majority and minority spin states: energy bands around Fermi level from about -5 to 4 eV are mainly related to -d-states(t2g) at Ti(1), Ti(2), and the Fe atoms, from -6 to -9 eV belong mainly to -s-and -p-states of the Y atom (Y=Fe, Ni), that are isolated from other groups. The presence of the majority spin states at the Fermi level and the presence of a band gap in the minority spin state confirm the HM nature of Ti_2YAl (Y=Fe, Ni) alloys. This gap in the HM band specially comes from the strong hybridization between the Ti(1)/Ti(2) atom and the Y atom (Y=Fe, Ni). The HM calculated band gap of Ti_2YAl (Y=Fe, Ni) alloys are listed in table 3 and compared with referenced values of literature.

Table 3. Calculated HM Band gap of Ti_2YAl (Y= Fe, Ni).

	Eg (eV)	
Alloy	**Calculated**	**Literature**
Ti_2FeAl	0.56	0.57795 [26] 0.555 [27] 0.53 [28]
Ti_2NiAl	0.44	0.28 [28] 0.46 [29] 0.57 [30]

IOP Conf. Series: Journal of Physics: Conf. Series **1081** (2018) 012008 doi:10.1088/1742-6596/1081/1/012008

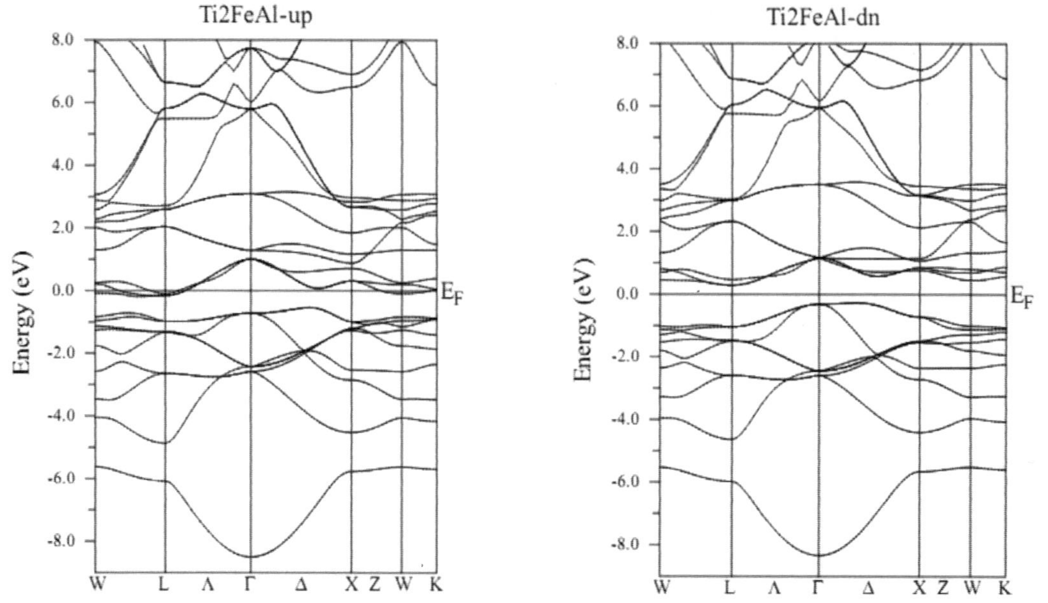

(a)

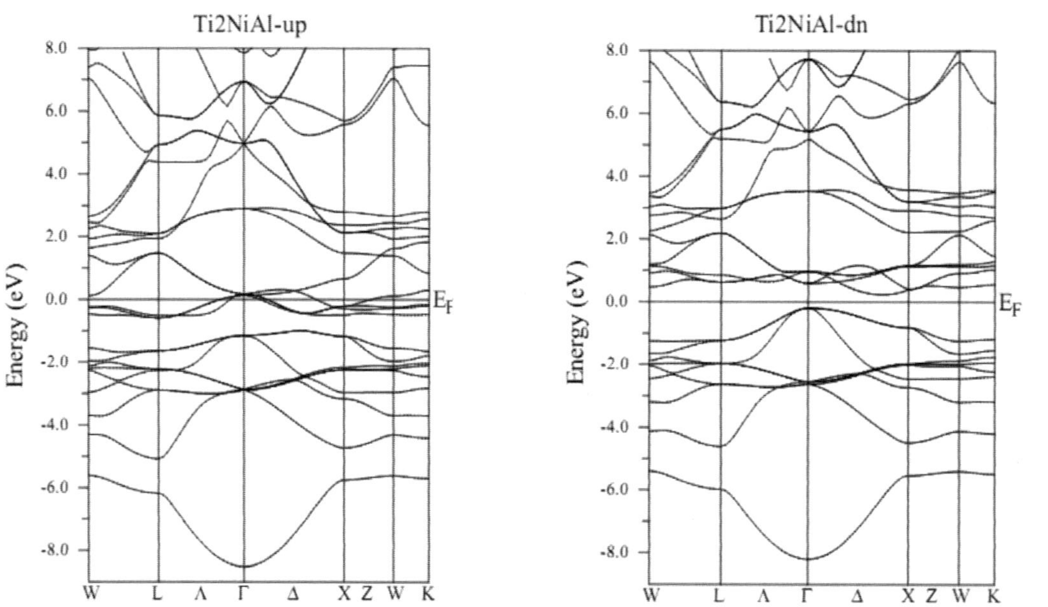

(b)

Figure 2. Spin resolved band structures in (a) Ti_2FeAl and (b) Ti_2NiAl alloys.

The calculated atomic magnetic moments are shown in table 4.The M_{tot} of these alloys consist in respective stationary values of $1\mu_B$ and $3\mu_B$ with respect to total valence electrons 19 and 21. This result is in good agreement with our expected results, making it consistent with the Slater-Pauling (SP) equation $M_{tot}=Z_{tot}-18$ in which Z_{tot} is the total number of valence electrons. M_{tot} in Ti_2YAl (Y=Fe, Ni) alloys contains four contributions: one from each of Ti(1), Ti(2), Y (Y = Fe, Ni) atoms and the interstitial region, and these values have shown that the main contribution to M_{tot} comes from the Ti atoms (1). The Ti(1) and Ti(2) atoms have positive magnetic moments while the Fe atom has a negative magnetic moment. M_{tot} of the atom Y (Y =Fe, Ni) has a very small value in all the three alloys.

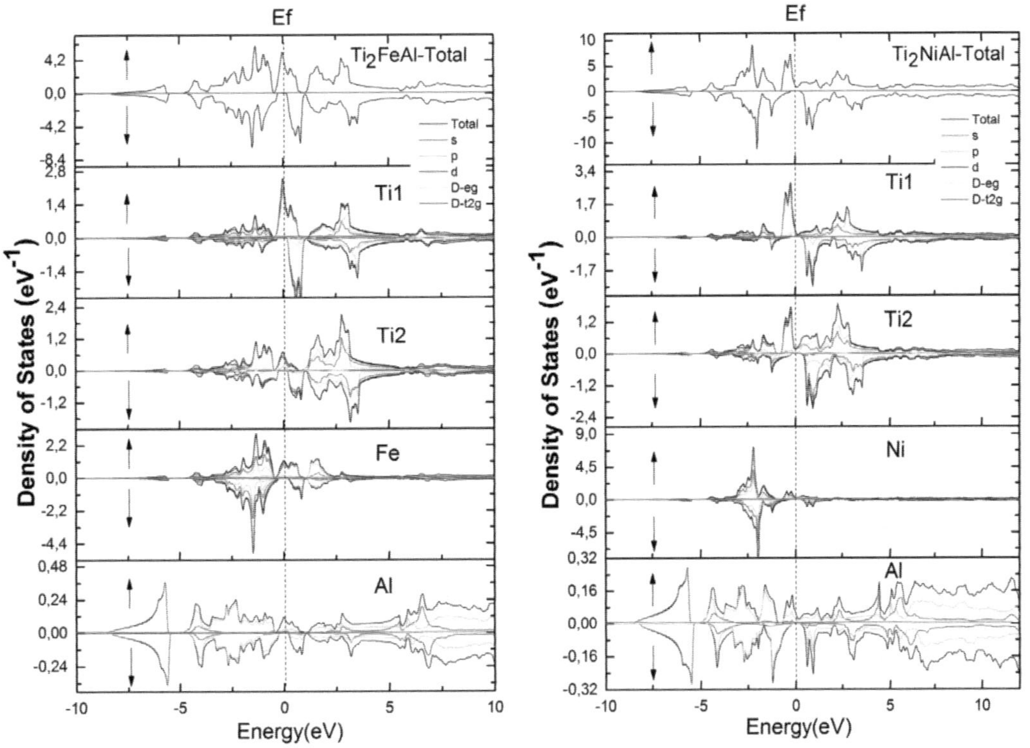

Figure 3. Spin-polarized total densities of states (DOS) and atom-projected DOS for the Ti_2FeAl (left graph) and Ti_2NiAl (right graph).

Table 4. Total and partial magnetic moments of Ti_2YAl (Y= Fe, Ni) compounds.

Compound	*M(Ti1)*	*M(Ti2)*	*M(Y)*	*M(Al)*	*Minterstitial*	M_{Tot}
Ti_2FeAl	0.82851	0.48816	-0.60998	0.01284	0.28100	1.00053
	0.86238 [26]	0.56666 [26]	-0.83923 [26]	0.00531 [26]	0.40475 [26]	0.99987 [26]
	0.885 [28]	0.620 [28]	-0.963 [28]	0.021 [28]		1.000 [28]
	1.21 [29]	0.84 [29]	-1.02 [29]	-0.02 [29]		1.00 [29]
	0.987 [27]	0.639 [27]	-0.758 [27]	0.027 [27]		0.981 [27]
Ti_2NiAl	1.36213	0.86233	0.16313	0.00901	0.60827	3.00487
	1.332[28]	0,843[28]	0.170[28]	0.020[28]		3.00[28]
	1.83 [29]	1.15[29]	0.10[29]	-0.08[29]		3.00[29]
						3.00[30]

4. Conclusions

In summary, we have studied the structural, electronic and magnetic properties of Ti_2FeAl and Ti_2NiAl alloys of the Hg_2CuTi-type structure. These alloys were found to be half-metallic ferrimagnets and to possess total magnetic moments 1 μ_B and 3 μ_B respectively. These results are in agreement with the Slater-Pauling rule.

References

[1] De Groot R A and Buschow K H J 1986 *J. Magn. Magn. Mater.* **54–57** 1377
[2] Wolf S A , Awschalom D D, Buhrman R A, Daughton J M, von Molnar S, Roukes M L, Chtchelkanova A Y and Treger D M 2001 *Science* **294** 1488
[3] Zutic I, Fabian J and Sharma S D 2004 *Rev. Mod. Phys.* **76** 323
[4] Chen X Q, Podloucky R and Rogl P 2006 *J. Appl. Phys.* **100** 113901
[5] Ozdogan K, Galanakis I, Sasıoglu E and Aktas B 2007 *Solid State Commun.* **142** 492
[6] Kandpal H C, Fecher G H and Felser C 2007 *J. Phys. D: Appl. Phys.* **40** 1507
[7] Liu G D, Dai X F, Lui H Y, Chen J L, Li Y X, Xiao G and Wu G H 2008 *Phys. Rev. B* **77** 14424
[8] Ozdogan K and Galanakis I 2009 *J. Magn. Magn. Mater.* **321** L34
[9] Sharma V, Solanki A K and Kashyap A 2010 *J. Magn. Magn. Mater.* **322** 2922
[10] Szotek Z, Temmerman W M, Svane A, Petit L, Stocks G M and Winter H 2004 *J. Magn. Magn. Mater.* **272** 1816
[11] Song W, Wang J and Wu Z 2009 *Chem. Phys. Lett.* **482** 246
[12] Lv S, Li H, Han D, Wu Z, Liu X and Meng J 2011 *J. Magn. Magn. Mater.* **323** 416
[13] Zhang Y, Liu W and Niu H 2008 *Solid State Commun.* **145** 590
[14] Saeed Y, Nazir S, Shaukat A and Reshak A H 2011 *J. Magn. Magn. Mater.* **322** 3214
[15] Galanakis I and Mavropoulos P 2003 *Phys. Rev.* B **67** 104417
[16] Xu Y Q, Liu B G and Pettifor D G 2003 *Physica* B **329–333** 1117
[17] Yao K L, Gao G Y, Liu Z L and Zhu L 2005 *Solid State Commun.* **133** 301
[18] Yao K L, Gao G Y, Liu Z L, Zhu L and Li Y L 2005 *Physica* B **366** 62
[19] Ge X F and Zhang Y M 2009 *J. Magn. Magn. Mater.* **321** 198
[20] Heusler Verh F 1903 *Deutsch. Phys. Ges.* **5** 219
[21] Xing N, Li H, Dong J, Long R and Zhang C 2008 *Comput. Mater. Sci.* **42** 600
[22] Luo H, Zhu Z, Liu G, Xu S, Wu G, Liu H, Qu J and Li Y 2008 *J. Magn. Magn. Mater.* **320** 421
[23] Xing N, Gong Y, Zhang W, Dong J and Li H 2009 *Comput. Mater. Sci.* **45** 489
[24] Yao Z, Gong S, Fu J, Zhang Y S, Yao K L 2010 *Solid State Commun.* **150** 2239
[25] Kanbur U and Gokoglu G 2011 *J. Magn. Magn. Mater.* **323** 1156
[26] Dahmane F, Benalia S, Djoudi L, Tadjer A, Khenata R, Doumi B and Aourag H 2015 *J Supercond Nov Magn.* **28** 3099
[27] Liping M, Yongfan S and Yu H 2014 *J. Magn. Magn.Mater.* **369** 205
[28] Nan Z and Yingjiu J 2012 *J. Magn. Magn. Mater.* **324** 3099
[29] Wei X P, Deng J B, Mao G Y, Chu S B and Hu X R 2011 arXiv:1110.5411v1
[30] Feng L, Tang C, Wang S and He W 2011 *J. Alloys Compd.* **509** 5187
[31] Andersen O K 1975 *Phys. Rev.* B **42** 3060
[32] Hohenberg P and Kohn W 1964 *Phys. Rev.* B **136** 864
[33] Kohn W and Sham L J 1965 *Phys. Rev.* **140** 1133
[34] Blaha P, Schwarz K, Madsen G K H, Kvasnicka D and Luitz J, 2008 WIEN2K, Augmented Plane Wave Plus Local Orbital's Program for Calculating Crystal Properties, University of Technology, *ISBN 3-9501031-1-2* Vienna
[35] Wu Z and Cohen R E 2006 *Phys. Rev.* B **73** 235116
[36] Murnaghan F D 1944 *Proc. Natl. Acad. Sci. USA* **30** 244

Electronic and elastic properties of Ba_2HgSn and Ca_2HgSn Rattling Heusler

Y Guermit[1,2], M Drief[1], N Benkhettou[1], T Lantri and D Rached[1]
[1] Laboratoire des Matériaux Magnétiques, Faculté des Sciences, Université Djillali Liabés de Sidi Bel-Abbés 22000, Algeria

E-mail: gyoucef75@gmail.com

Abstract. Semiconductor electronic components are the active elements of integrated circuits, and they are at the origin of all advances in electronics. A new semiconductor class consists in Rattling Heusler. The purpose of this work is to study the structural, electronic and elastic properties of X_2HgSn (X=Ba, Ca) Heusler alloys by using the first-principle projector augmented wave potential within the generalized gradient approximation. Our calculations show that these compounds are all semiconductor and are ductile manner.

1. Introduction

Currently, the new technology seeks its raw material in the elements periodic table, based on the fact that the combination of two different materials does not present a combination of their properties but rather gives rise to new specific features to the formed alloy. This motivation was a driving force of the development of materials science and engineering as an important branch like chemistry, physics and metallurgy. Materials science aims to determining the relationships between structures and material properties, while materials engineering is focused on the development and the design of materials with known structures and properties. Heusler X_2YZ alloys are among the materials that are attracting growing interest since the pioneer discovery, made by the German chemist Friedrich Heusler in 1903, of ferromagnetism in Cu-Mn-Al alloys in which the constituent elements are not ferromagnetic [1]. Indeed, compounds of Heusler form present a remarkable class of materials due to their various physical properties and potential applications. Whereas, the properties of these X_2YZ compounds, with alkaline earth elements (Ba,Sr, Ca) in the X sublattice, whereas Y are noble metals (Au, Hg) and Z are main group elements (Sn, Pb, As,Sb, Bi) [2], have not yet been reported in literature. In this paper, we present a detailed study on the site preference, electronic and elastic properties of Heusler alloys X_2HgSn (X=Ba, Ca) using first-principles calculations.

2. Computational Method

The first-principles calculations are performed using the Full-Potential Linearized Augmented Plane Wave (FP-LAPW) method as implemented in WIEN2K code [3]. The exchange-correlation potential

[2] To whom any correspondence should be addressed.

IOP Conf. Series: Journal of Physics: Conf. Series **1081** (2018) 012009 doi:10.1088/1742-6596/1081/1/012009

is calculated within the Generalized Gradient Approximation (GGA) by Perdew Burke Ernzerhof (PBE) [4] is also applied in calculations. In this method, no shape approximation on either potential or the electronic charge density is made. Using this method, valence and core states origin energy have been separate, and –6 Ryd energy bound separated between valence electrons and core states was chosen. In the calculations reported here, we use an RMT K_{max}=8 parameter, which determines matrix size (convergence), where K_{max} is the plane wave cut-off and RMT is the smallest atomic sphere radius. The density of states is calculated by the histogram method and the position of the Fermi level is found by integrating consistent calculations. The energy convergence criterion for the self-consistent field calculations was set to 10^{-4} Ryd, whereas the charge convergence was set to 10^{-4}. The Fermi level energy was adjusted to zero, and we used 13 × 13× 13 meshes which represent 84 k points in the first Brillouin zone integrations in reciprocal space Ire performed using the special points method.

3. Results and discussion

3.1. Structural stability

Heusler alloys X_2YZ (Cu_2MnAl-type) crystallize in the Fm3m (N 225) regular cubic space group [5, 6], where the X atoms occupy the 8c Wyckoff position (1/4, 1/4, 1/4) and the Y and Z atoms are respectively at the 4a (0, 0, 0) and 4b (1/2, 1/2, 1/2) positions (see figure 1(a)). The crystal can be crystallized in F43m (N 216) inverse cubic structure (Hg_2CuTi-type) [7], the X atoms are placed on the positions 4a (0, 0, 0) and 3d (3/4, 3/4, 3/4) while the atoms Y and Z are respectively located at 4b (1/2, 1/2, 1/2) and 4c (1/4, 1/4, 1/4) as shown in figure 1(b).

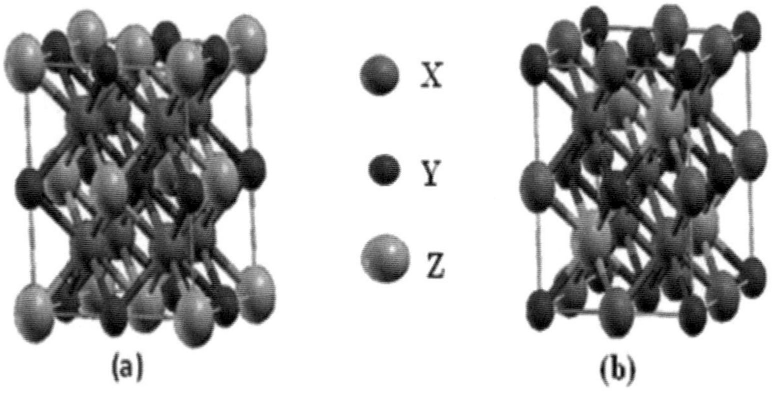

Figure 1. Different types of Heusler structures: (a) regular-Heusler, (b) inverse-Heusler.

We have calculated the total energy as a function of lattice constant of X_2HgSn (X=Ca, Ba) compounds, in the regular cubic and the inverse cubic structures, for the ferromagnetic (Ferro) and nonmagnetic (NM) states. The plot of calculated total energies versus reduced volume for compounds Ca_2HgSn and Ba_2HgSn are given in figure 2 and figure 3 respectively. The total energies versus changed volumes have been fitted to Murnaghan's equation of state [8] in order to determine the ground state properties, such as equilibrium lattice constant a, bulk modulus B and its pressure derivative B′. The calculated structural parameters of these compounds are reported in table 1.

The results obtained with the GGA-PBE method shows that the most stable structure for all compounds is Cu_2MnAl-type in the nonmagnetic phase.

Table 1. Calculated lattice constants a, bulk modulus B, first derivatives B' and equilibrium energies E for Ba_2HgSn and Ca_2HgSn compounds.

Alloy	Type	Phase	a (Å)	B (GPa)	B'	E (eV)
Ba_2HgSn	Cu_2MnAl	FM	8.48	25.36	4.29	-84239.640248
		NM	8.48	25.35	4.27	-84239.640250
	Hg_2CuTi	FM	8.31	23.20	4.46	-84239.537754
		NM	8.31	23.29	4.51	-84239.537761
Ca_2HgSn	Cu_2MnAl	FM	7.68	36.07	4.14	-54403.727262
		NM	7.68	36.11	4.50	-54403.727271
	Hg_2CuTi	FM	7.75	30.56	3.88	-54403.609839
		NM	7.75	30.40	4.21	-54403.609802

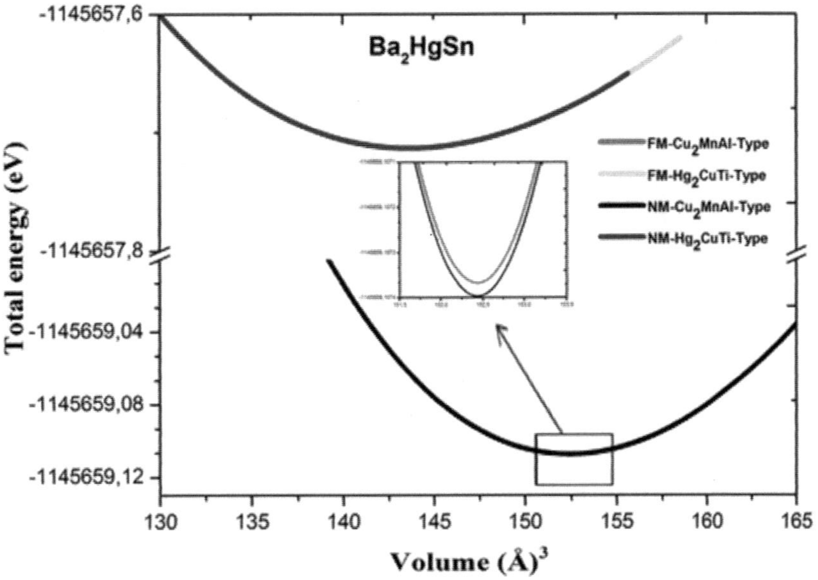

Figure 2. Calculated total energies as functions of volume for Ba_2HgSn Heusler alloys.

IOP Conf. Series: Journal of Physics: Conf. Series **1081** (2018) 012009 doi:10.1088/1742-6596/1081/1/012009

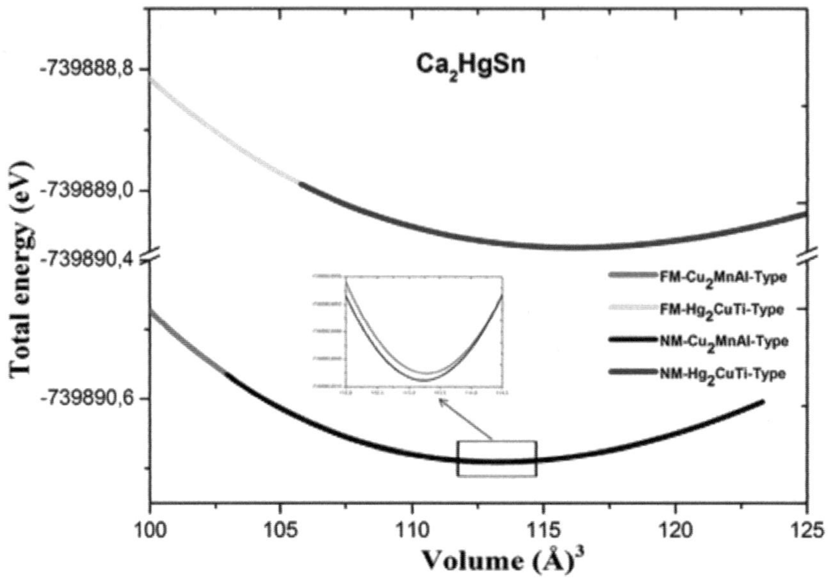

Figure 3. Calculated total energies as functions of volume for Ca_2HgSn Heusler alloys.

3.2. Electronic properties

The calculated band structures of Heusler alloys X_2HgSn (X=Ba, Ca) at equilibrium lattice constant along the high symmetry lines of the BZ are presented in figure 4. This figure shows clearly the semiconducting behaviour of Ba_2HgSn and Ca_2HgSn which are drawn along the symmetry directions in the first Brillouin zone. Calculations predict the valence band maximum at between L-Λ and the conduction band minimum at Δ point resulting in an indirect band gap for the Ba_2HgSn alloy. In contrast, the Ca_2HgSn compound exhibits direct band gap (L-L).

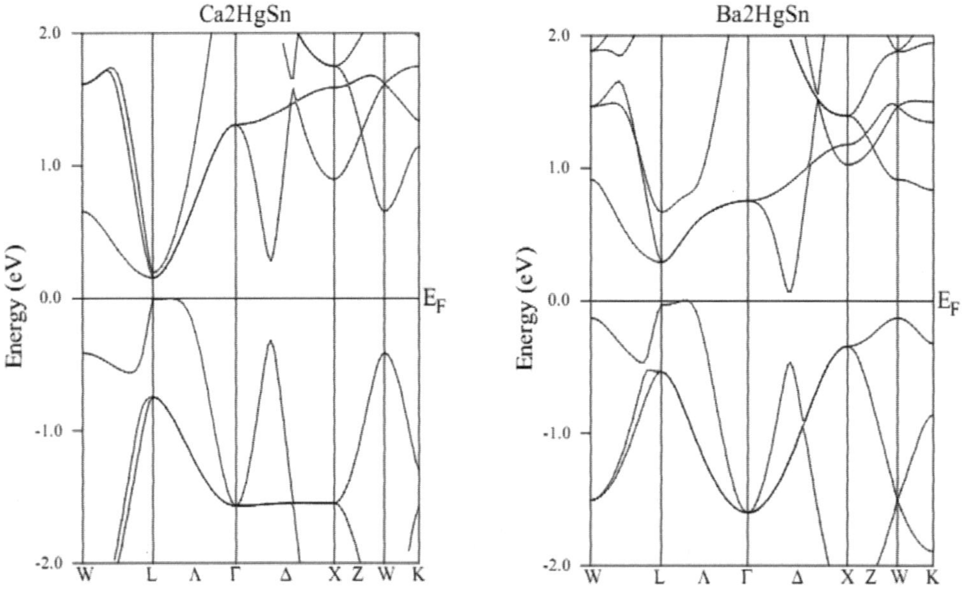

Figure 4. Band structures of Ca_2HgSn (left graph) and Ba_2HgSn (right graph) Heusler alloys.

3.3. Elastic properties

The elasticity of a solid body is its response in the form of slight deformation when subjected to external mechanical stresses. The stresses are described by tensors which determine the direction of forces and the plane on which they are applied.

The numerical calculation of the elastic constants for the studied alloys can be performed by the method of Thomas Charpin implemented in the code WIEN2k.In cubic crystal, there are only three independent elastic constants, namely, C_{11}, C_{12} and C_{44}. By calculating the total energy as a function of strain, these elastic constants can be determined [9]. The conditions of elastic stability in cubic crystal are $C_{11} - C_{12} > 0$, $C_{11} > 0$, $C_{44} > 0$, $C_{11} + 2C_{12} > 0$, $C_{12} < B < C_{11}$ [10]. Besides, other elastic parameters can be calculated such as the Shear modulus G, Young's modulus E, Poisson's ratio ν and Anisotropy factor A of X_2HgSn (X= Ba, Ca) using the Voigt–Reuss–Hill approximations [11]:

$$G = \frac{1}{2}\left(\frac{C_{11} - C_{12} - 3C_{44}}{5} + \frac{5C_{44}(C_{11} - C_{12})}{4C_{44} + 3(C_{11} - C_{12})}\right) \quad (1)$$

$$E = \frac{9BG}{G + 3B} \quad (2)$$

$$\nu = \frac{3B - 2G}{2(3B + G)} \quad (3)$$

$$A = \frac{3C_{44}}{C_{11} - C_{12}} \quad (4)$$

Our results are presented in table 2. One should note that no experimental values or theoretical calculations for the elastic constants of Ba_2HgSn and Ca_2HgSn have been reported before. According to the stability criteria, the studied compounds with GGA are mechanically stable because the value ($C_{11} - C_{12}$) is positive for all the compounds. Our calculated values using GGA of B/G are 1.74, 1.67 for Ba_2HgSn and Ca_2HgSn respectively and $\nu > 1/3$ for all compounds. We can conclude the studied components are ductile manner according to [12,13].

Table 2. Calculated elastic constants: Anisotropy factor A, Shear modulus G, Young's modulus E and the Poisson's ratio ν of Ba_2HgSn and Ca_2HgSn.

Alloy	C_{11}	C_{12}	C_{44}	G	E	ν	B/G	A
Ba_2HgSn	43.0	16.579	15.6	14.6	36.8	0.26	1.74	1.18
Ca_2HgSn	59.5	24.5	25.1	21.7	54.8	0.25	1.67	1.43

4. Conclusion

In summary, we have performed first principles calculations based on the FP-LAPW method within the GGA. We have studied structural, electronic and elastic properties of Ba_2HgSn and Ca_2HgSn full Heusler alloy. The study shows that the most stable structure for all compounds is Cu_2MnAl-type in the nonmagnetic phase. The characterization of electronic properties shows that the compounds Ba_2HgSn and Ca_2HgSn are semiconductor materials. The elastic constants C_{11}, C_{12} and C_{44} of these alloys were calculated for the first time, and the obtained results could provide a basic reference for other similar work.

References

[1] Heusler Verh F. 1903 *Deutsch. Phys. Ges.* **5** 219
[2] He J 2016 arXiv:1604.03827v2
[3] Blaha P, Schwarz K, Madsen G K H, Kvasnicka D and Luitz J, 2008 WIEN2K, Augmented Plane Wave Plus Local Orbital's Program for Calculating Crystal Properties, University of Technology, ISBN 3-9501031-1-2 Vienna, Austria
[4] Perdew J P, Burke K and Ernzerhof M 1996 *Phys. Rev. Lett.* **77** 3865
[5] Heusler O 1934 *Ann. Phys.* **155** 411
[6] De Groot R A, Mueller F M, van Engen P G and Buschow K H J 1983 *Phys. Rev. Lett.* **50** 2024
[7] Puselj M and Ban Z 1969 *Croat. Chem. Acta.* **41** 79
[8] Murnaghan F D 1944 *Proc. Natl. Acad. Sci. USA* **30** 244
[9] Mehl M J, Osburn J E, Papaconstantopoulos D A and Klein B M 1990 *Phys. Rev.* B **41** 103112
[10] Sin'ko G V and Smirnov N A 2002 J. Phys. Condens. Matter **14** 6989
[11] Hill R1952 *Proc. Phys. Soc. Lond.* A **65** 349
[12] Pugh S F 1954 *Philos. Mag.* **45** 823
[13] Frantsevich I N, Voronov F F and Bokuta S A 1983 Elastic Constants and Elastic Moduli of Metals and Insulators Handbook, ed. I. N. Frantsevich, Kiev

Dynamics of the etching effects on the optoelectronic properties of ZnO nanorods

I Boukhoubza[1,2,4], M Khenfouch[2,3], M Achehboune[1,2], B Mouthudi[3], I Zorkani[1], A Jorio[1]

[1] University Sidi Mohammed Ben Abdellah, Faculty of Sciences Dhar el Mahraz, Laboratory of Solid state Physics, Group of Nanomaterials and Renewable Energies, PO Box 1796 Atlas Fez 30 000, Morocco

[2]Africa Graphene Center, Physics department, Eureka building, College of Science, Engineering and Technology, Science Campus, University of South Africa, Cnr Christiaan de Wet & Pioneer Avenue Florida 1709, Johannesburg, South Africa.

[3] UNISA university of South Africa, Department of Physics, College of Science, Engineering and Technology, Science Campus, Cnr Christiaan de Wet & Pioneer Avenue Florida 1709, Johannesburg, South Africa

E-mail: boukhoubza.issam00@gmail.com

Abstract. Zinc oxide (ZnO) is a very promising material for optoelectronic applications thanks to its wide direct band gap (3.4 eV), and a high exciton binding energy (60 meV). ZnO based nanostructures are presently being explored for a wide range of applications in nanolasers, nanogenerators, gas sensors, light emitting diodes and solar cells. In this study, we report the preparation of ZnO nanorods and the dynamics of their chemical etching under different concentrations. Characterization was carried out using X-ray diffraction, scanning electronic microscopy and photoluminescence. Thus, the study provides an investigation of the etching time and concentration effects on the optoelectronic properties. We found that all etching conditions are causing a remarkable proportional changes on the materials photonics as it is mainly changing the surface and creating defects. Hence, this study exhibits distinct advantages for optoelectronic devices.

1. Introduction

Zinc oxide (ZnO) is one of the most important semiconductors. Indeed, ZnO has a wide band gap (Eg=3.37eV), a direct band gap structure, a large exciton binding energy (60 meV), excellent chemical and thermal stability at room temperature [1]. In recent years, several publications considered one dimensional ZnO nanostructures such as nanorods [2], nanowires [3], nanobelts [4], nanotubes [5], and nanorings [6] etc. These materials have attracted great interest due to their promising applications in electronic, optoelectronic and sensing devices [7]. In particular, ZnO nanorods present excellent crystalline quality, high surface to volume ratio and high electron mobility [8]. Among these, ZnO nanorods have been extensively used in the recent past years due to their multifunctional device

[4] To whom any correspondence should be addressed

applications in the areas of ultraviolet lasers [9], light emitting diodes (LEDs) [10], nanogenerators [11,12], field emission devices [13,14], gas sensor [15], and dye sensitized solar cell [16]. Presently, various methods have been developed to fabricate ZnO nanorods, such as physical vapor deposition (PVD) [17], chemical vapor deposition (CVD) [18], metal organic chemical vapor deposition (MOCVD) [19], molecular beam epitaxy (MBE) [20], pulsed laser deposition (PLD) [21], hydrothermal synthesis [22], chemical bath deposition [23], and chemical etching method [24].
Among all these techniques, chemical etching method is a low cost and environmentally friendly technique. This technique has been applied in industry over past decades. Wet chemical etching has great advantages due to its easy operation, time-saving, and capacity of anisotropic/isotropic etching [25]. Although the etching parameters consist of the etching time and the concentration of the etching solution, it is interesting to note that the etching rate also depends on the crystallographic orientation of the ZnO [26, 27]. Few studies discussed the effects of the etching parameters on the sizes and morphologies of ZnO nanorods. For example, Lei et al [28] demonstrated feasibility of the growth of ZnO nanorods by directly etching the substrate. Gan et al [7] electrodeposited well-aligned ZnO nanorods on ITO glass and analyzed the effects of both the etching periods and KOH concentration on the morphology of the ZnO nanotubes. Gruzintsev et al. [29] reported the effect of chemical etching on the luminescent properties of ZnOnanorods.
In this paper, ZnO nanorods were prepared by a chemical etching method under different concentrations. The concentration of the etching solution and the etching time were regulated respectively and their effects on the morphology of the ZnO nanorods were also discussed in details. The samples were then characterized by X-ray diffraction (XRD), Scanning Electron Microscopy (SEM) and photoluminescence (PL) to investigate their structural and optical properties.

2. Experimental method

ZnO precursors used in this work were purchased from Sigma-Aldrich. To synthesize ZnO nanorods, an equi-molar aqueous solution of 0.1 M zinc nitrate (Zn(NO3)2) and hexamethylenetetramine ($C_6H_{12}N_4$, HMT) was prepared using deionized water. This mixture was heated at a constant temperature of 90°C in an oven for 24 h. The obtained materials were deposited on Si ⟨100⟩ substrates and dried at 50°C for 2 hours. Then, the formed films were etched with solutions of NaOH of different concentrations.

3. Results and discussion

3.1. Structural properties

Figure 1 shows the XRD pattern of pure ZnO and ZnO nanorods prepared at different concentrations by the chemical etching method. All the diffraction peaks matched greatly well with hexagonal phase without getting any additional peaks except the peaks which are characteristic of ZnO. The intense peaks in the XRD pattern of the samples clearly showed the formation of hexagonal wurtzite phase of ZnO corresponding to (101) reflection plane in all samples, which is the preferred orientation. In addition, the intensity of the peaks (100) and (002) decreased with increasing etching concentration, which can be due to the incorporating of impurities in the lattice of ZnO. Also, the reduction in the intensity of XRD spectra is observed with the increasing concentration, this can be attributed to the dissolution of (100), (002) and (101) during the etching process.

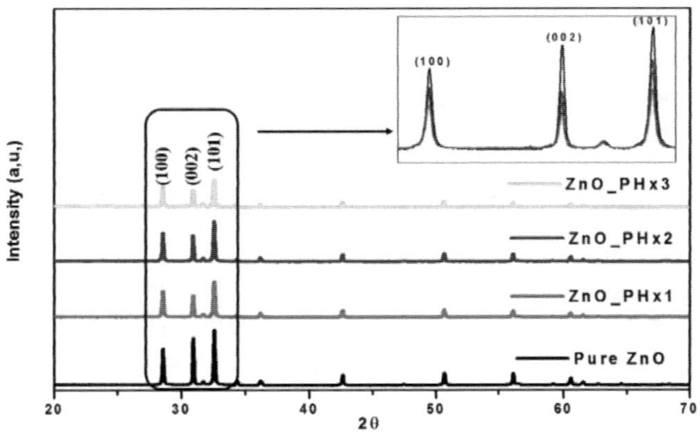

Figure 1. XRD patterns of the pure ZnO and ZnO nanorods etched at different concentrations.

Figure 2 shows the XRD pattern of pure ZnO and ZnO nanorods prepared with different etching times 30 min, 60min, and 120min, respectively. All the diffraction peaks matched very well with the hexagonal wurtzite phase of ZnO and no other crystalline phases are detected. Also, the high intensity (101) peak in the XRD pattern compared to (100) and (002) peaks, corresponding to the reflection plane in all samples, which is the preferred orientation. The intensity of the (101) peak of the ZnO nanorods decreased with increasing etching time from 30 min to 60 min, and then increased with further increasing etching time to 120 min. This can be explained by the reduction of crystallinity following directions (100) and (002).

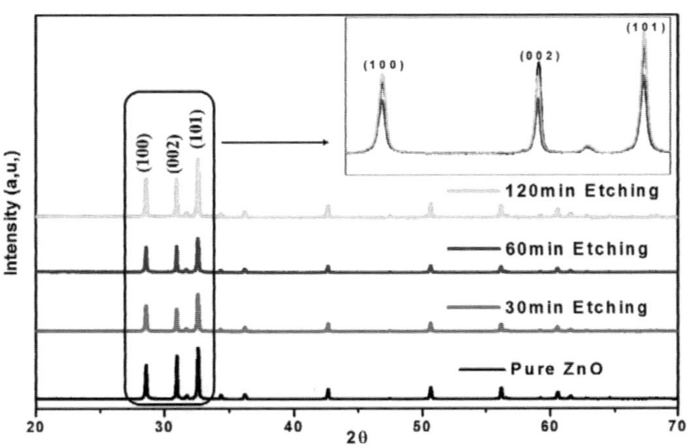

Figure 2. XRD patterns of the pure ZnO and ZnO nanorods etched with different etching times.

Figure 3 shows SEM images of pure ZnO nanorods and etched by the NaOH solution for 60 min. From the image (a), it can be observed that the ZnO nanorods with high density and well-defined hexagonal cross sections were grown vertically. The average diameter of the nanorods was about 400 nm and the surface of ZnO nanorods was very smooth. After etching, these nanorods for 60 min, large scale hexagonal ZnO nanorods preserve the same shape and diameter as before. However, image (b) shows clearly several defects on the surface of ZnO nanorods after etching treatment.

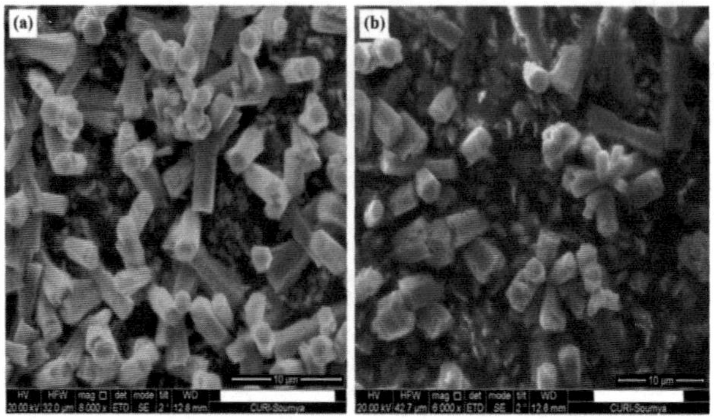

Figure 3. SEM images of zinc oxide nanorods: (a) as-grown, (b) after etching.

3.2. Optical properties

The PL spectrum was measured to analyze the optical properties of ZnO nanorods and defect types of the products which are important for optoelectronic applications. Figure 4 below presents PL spectra of the as-grown ZnO and ZnO nanorods prepared by the hydrothermal method. All the PL spectra show a distinct emission located around 560 nm. As can be seen in this figure, the intensity of the PL spectra is initially increasing as a function of the etching concentration. The differences in the PL band-edge emission peaks could be attributed to the increasing concentration of defects in the ZnO nanorods. A higher broad green-band emission peak is owing to the attendance of a number of ionized oxygen vacancy (V_0^+) defects. Guo et al. have found that the oxygen vacancy results from the concurrence among O atoms entering the ZnO lattice and those exciting this lattice by evaporation [30].

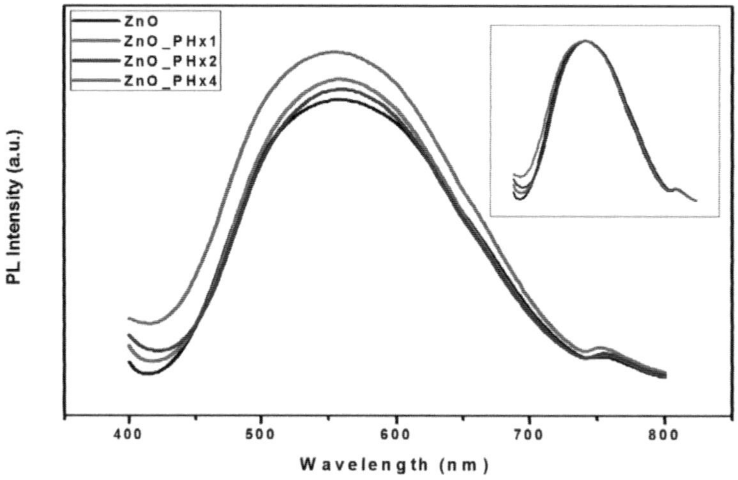

Figure 4. PL spectra of the as-grown ZnO and ZnO samples etched at various concentrations of NaOH solution.

Figure 5 shows PL spectra of the as-grown ZnO and ZnO nanorods etched for different times. The peaks shown in this figure are the PL band-edge emission peaks of ZnO located at approximately 560 nm and are commonly attributed to oxygen vacancies [31]. Up to 60 min, the intensity of the emission peaks increases with increasing the etching time. However, as the samples were etched over longer periods of time, the intensity of the PL spectra starts to decrease. As viewed in this figure, the intensity of the green band emission peak decreases (to 120 min). This can be explained by the fact that the effect of etching has a saturation on the emission. In other words, to increase the intensity of the peaks, the etching should not exceed a given time. Another explanation could be that the intensity of the PL band emission peak, which decreases after 120 min, is awarded to the decrease in the surface to volume ratio of the ZnO nanorods that conducted the decrease of the PL band edge intensity which is in good agreement with previous work [32].

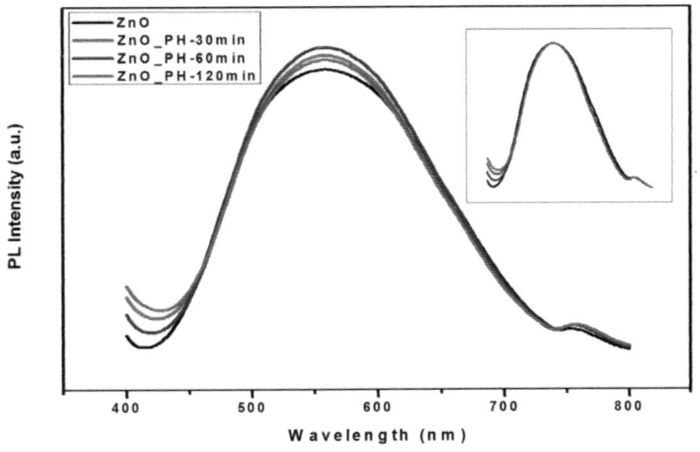

Figure 5. PL spectra of as-grown ZnO and ZnO nanorods etched by NaOH solution for different times.

4. Conclusion

In summary, we have synthesized ZnO nanorods by the hydrothermal method and etched them chemically. The analysis of X-ray diffraction showed that the samples had hexagonal wurtzite structure with (101) as a preferred orientation for the growth of the ZnO nanorods. After etching, SEM images also showed some defects in the samples which lead to a change in the surface morphology of the ZnO nanorods. Furthermore, PL spectra showed that the peaks intensity increases with increasing etching time. These results may be explained by the decrease in the surface to volume ratio attributed to the longer etching time. Therefore, this process can be used to enhance the efficiency of some optoelectronic devices.

Acknowledgments

This work was supported bythe University Sidi Mohammed Ben Abdellah USMBA (Morocco), University of South Africa Department of Physics and Africa Graphene Center (South Africa).

References

[1] Hatamie A, Khan A,Golabi M,Turner A P F,Beni V, Mak W C, Sadollahkhani A, Alnoor H, Zargar B, Bano S,Nur N and Willander M 2015 *Langmuir* **31** 10913–10921
[2] Shouli B, Liangyuan C, Dianqing L, Wensheng Y, Pengcheng Y, Zhiyong L, Aifan C and Liu C C 2010 *Sensors and Actuators B* **146** 129–137
[3] Yang P, Yan H, Mao S, Russo R, Johnson J, Saykally R, Morris N, Pham J, He R and Choi H J

2002 *Adv. Funct. Mater.* **12** 323–331
[4] Huang Y, He J, Zhang Y, Dai Y, Gu Y, Wang S and Zhou C 2006 *J. Mater. Sci.* ***41*** 3057–3062
[5] Xi Y, Song J, Xu S, Yang R, Gao Z, Hu C and Wang Z L 2009 *J. Mater. Chem.* **19** 9260–9264
[6] Kong X Y, Ding Y, Yang R and Wang Z L 2004 *Science* **303** 1348–1351
[7] Gan X, Li X, Gao X and Yu W 2009 *Journal of Alloys and Compounds* **481** 397–401
[8] Nandi R and Majo S S 2017 *Applied Surface Science* **399** 305–312
[9] Huang M H, Mao S, Feick H, Yan H, Wu Y, Kind H, Weber E, Russo R and Yang P 2001 *Science* **292** 1897–1899
[10] Park W I, and Yi G C 2004 *Adv. Mater.* **16** 87–90.
[11] Buchine B A, Hughes W L, Degertekin F L and Wang Z L 2006 *Nano Lett.* **6** 1155–1159.
[12] Xu S, Qin Y, Xu C, Wei Y, Yang R and Wang Z L 2010 *Nat. Nanotechnol.* **5** 366–373.
[13] Zhu Y W, Zhang H Z, Sun X C, Feng S Q, Xu J, Zhao Q, Wang R M and Yu D P 2003 *Appl. Phys. Lett.* **83** 144– 146.
[14] Wang W Z, Zeng B Q, Yang J, Poudel B, Huang J Y, Naughton M J and Ren Z F 2006 *Adv. Mater.* **18** 3275–3278.
[15] Kumar R, Al-Dossary O, Kumar G and Umar A 2015 *Nano-Micro Lett.* **7** 97–120
[16] Valls I G and Cantu M L 2010 *Energy Environ. Sci.* **3** 789–795
[17] Xu S, Qin Y, Xu C, Wei Y, Yang R and Wang Z L 2010 *Nat.Nanotechnol.* **5** 366–373
[18] Cong G W, Wei H Y, Zhang P F, Peng W Q, Wu J J and Wang Z G 2005 *Appl. Phys. Lett.* **87** 231903–231906
[19] Park W I, Kim D H, Jung S W and Yi G C 2002 *Appl. Phys.Lett.***80** 4232–4234.
[20] Heo Y W, Varadarajan V, Kaufman M, Kim K, Norton D P, Ren F and Fleming P H 2002 *Appl. Phys. Lett.***81** 3046–3048.
[21] Sun Y, Fuge G M and Ashfold M N 2004 *Chem. Phys. Lett.* **396** 21–26.
[22] Wei H, Wu Y, Lun N and Hu C 2005 *Materials Science and Engineering* **393** 80–82
[23] Wu J, and Liu S C, 2002 *Adv. Mater.* **14** 215–218.
[24] Han X G, Jiang Y Q, Xie S F, Kuang Q, Zhou X, Cai D P, Xie X Z and Zheng L S 2010 *J. Phys. Chem.* **114**10114–10118
[25] Jiang Q, Lu J, Zhang J, Yuan Y, Cai H, Wu C, Sun R, Lu B, Pan X and Ye Z 2014 *Solar Energy Materials & Solar Cells* **130** 264–271
[26] Ching C G, Ooi P K, Ng S S, Ahmad M A, Hassan Z, Abu Hassan H and Abdullah M J 2013 *Materials Science in Semiconductor Processing* **16** 70–76
[27] Yoo D J, Nam S H, Kim M H, Jeong S H, Jee H G, Lee H G, Lee N E , Hong B Y, Kim Y G, Jung D J and Boo J H 2008 *Surface & Coatings Technology***202** 5476–5479
[28] Lei Y, Wang J, Qu F, Li H and Wu X 2012 *Micro & Nano Letters* **7** 485–488
[29] Gruzintsev A N, Red'kin A N, Yakimov E E and Barthou C 2008 *Inorganic Materials* **44** 1080–1085
[30] Guo M, Diao P and Cai S 2005 *J. Solid State Chem.***178** 1864–1873
[31] Greene L E, Law M, Goldberger J, Kim F, Johnson J C, Zhang Y, Saykally and Yang P D 2003 *Angew. Chem.* **42** 3031–3034.
[32] Ching C G, Lu L, Ang C I, Ooi P K , Ng S S, Hassan Z and Abu Hassan H 2013 *Sains Malays.* **42** 1327–1332

IOP Conf. Series: Journal of Physics: Conf. Series **1081** (2018) 012011 doi:10.1088/1742-6596/1081/1/012011

Band gap behavior of thallium phosphide and gallium phosphide binary compounds and $Ga_x\,Tl_{1-x}\,P$ ternary alloys

M Merabet[1,2,3], **L Djoudi**[1,2], **M Sayah**[2], **S Benalia**[1,2] **and D Rached**[1]
[1] Laboratoire des Matériaux Magnétiques, Faculté des Sciences, Université Djillali Liabés de Sidi Bel-Abbés 22000, Algeria
[2] bUniversity of Tissemsilt, Institute of Science and Technology 38000, Algeria

E-mail: merabetmostefa@yahoo.fr

Abstract. Full-potential linear muffin-tin orbitals method has been implanted within density functional theory and within local density approximation. The structural and electronic properties of GaP and TlP binary compounds and their ternary alloys $Ga_x\,Tl_{1-x}\,P$ (x=0,25, 05 or 0.75) in the zinc-blende structure were investigated and calculations obtained results gave reasonable agreements with other published data.

1. Introduction

III-V semiconductor compounds have been extensively studied theoretically and experimentally during the last half century. Over many years, this family of semiconductors has demonstrated its potential for applications in optoelectronic devices (light-emitting diodes, laser diode, gas captors, solar cells), and devices in hostile environments (high temperatures, high frequency) with high performances [1–17].

The III–phosphide binary compounds appear as promising materials among the family of III–V semiconductors. At ambient conditions, these compounds crystallize in the zinc-blende structure and have an indirect band gap [18–32] apart from InP [33]. However, the weak optical transition in indirect gap materials, such as GaP, is a major limitation for their use in optical devices. The radiative transition in these materials is accompanied by phonon emission and the oscillator strength of such transitions is, therefore, inferior compared to the direct band-gap semiconductors such as GaAs [34].

Thus, the investigation about new materials, allowing the design and the construction of new optoelectronic devices, motivates our work with the aim of proposing new materials. These new semiconductor alloys exhibit a range of unexpected characteristics. Particularly, We have aimed to combine GaP to TlP since they possess different structural and electronic properties, in order to obtain new materials consisting in $Ga_x\,Tl_{1-x}\,P$ ternary alloys with intermediate properties.

It is expected that TlP crystallizes most probably in in the zinc-blende (B3), with a semi-metallic behavior in this phase [35]. However, GaP may have an indirect band gap in the zinc blende (B3) phase [36].

[3] To whom any correspondence should be addressed.

IOP Conf. Series: Journal of Physics: Conf. Series **1081** (2018) 012011 doi:10.1088/1742-6596/1081/1/012011

In order to take full advantage of the properties of these compounds for possible technological applications, a theoretical investigation of the structural and electronic properties is necessary. For this purpose, we used the first-principles method to investigate the structural and electronic properties of the binary compounds GaP and TlP and their ternary alloys $Ga_xTl_{1-x}P$ (x=0,25, 05 or 0.75) in the zinc-blende structure, using the state of the art Full-Potential Linear Muffin-Tin Orbital (FP-LMTO) method, in the framework of the Density Functional Theory (DFT), within the Local Density Approximation (LDA) for the exchange correlation functional.
In the following, we briefly describe the used method and the detailed calculations, and results of the structural and electronic properties of the studied binary and ternary compounds are then presented and discussed.

2. Calculation methodology

The calculations reported in this work were carried out using the FP-LMTO method [37, 38]. It consists in an implementation of the DFT, which is a universal quantum mechanical approach for many-body problems. In the latter, the quantum many-body problems of interacting electrons and nuclei can be mapped onto a system of one-electron equations called Kohn-Sham equations [39, 40]. In this method, the space is divided into an interstitial region (IR) and non-overlapping (MT) spheres centered at the atomic sites. In the IR region, the basis set consists of plane waves. Inside the MT spheres, the basis set is described by radical solutions of the one particle Schrodinger equation (at fixed energy) and their energy derivatives multiplied by spherical harmonics.
In order to achieve energy eigenvalues convergence, the charge density and potential inside the muffin-tin spheres are represented by spherical harmonics up to l_{max}=6. The exchange-correlation potential is treated within the local density approximation developed by Perdew–Wang [41]. The self-consistent calculations are considered to be converging when the total energy of the system is stable within 10-5 Ry. The k integration over the Brillouin zone is performed using the tetrahedron method [42].

3. Results and discussion

3.1. Structural properties

The majority of III-V compound semiconductor materials have a zinc blende crystal lattice structure. The zinc blende structure consists of two interpenetrating face centered cubic (FCC) sub-lattices, one containing group III atoms and the other containing group V atoms. The two sub-lattices are shifted relative to each other by (a/4,a/4,a/4) where a is the cubic lattice constant. In this subsection, we calculate the structural properties of the compounds GaP and TlP and their alloys in the cubic structure. The considered semiconductor ternary alloys are of the type $B_xA_{1-x}C$ having the basic cubic cell as the unit cell, in which there are four C anions and three A and one B, two A and two B, or one A and three B cations (respectively for x = 0.25, 0.50 or 0.75). We have started our FP-LMTO calculations of the structural properties with the zinc-blende structure and let the calculation forces to move the atoms to their equilibrium positions.
For the considered structures, we perform the structural optimization by calculating the total energies for different volumes around the equilibrium cell volume V_0 of the binary GaP and TlP compounds and their alloys. The calculated total energies are fitted to the Birch's equation of state [43] to determine the ground state properties such as the equilibrium lattice constant a_0, the bulk modulus B_0 and its pressure derivative B'_0. The calculated equilibrium parameters (a_0, B_0 and B'_0 are given in table 1). We remark that our results for GaP and TlP agree well with literature values; however, comparison was not always possible for ternary alloys phase because of the lack of data in the literature.

3.2. Electronic properties

The calculated band structure energies of binary compounds as well as their alloys using the FP-LMTO method within the LDA approximation along the higher symmetry directions of Brillouin zone are shown in figures 1and 2. From these figures, we remark an indirect band gap at the Γ-X point for GaP and a negative band gap for TlP. For $Ga_{0.75}Tl_{0.25}P$ and $Ga_{0.5}Tl_{0.5}P$, both valence band maximum (VBM) and the conduction Band minimum (CBM) are located at the Γ point, predicting a direct transition. Nevertheless, for $Ga_{0.25}Tl_{0.75}P$, one can clearly see the semi metallic nature of this compound. The important features of the band structure (main band gaps) are given in table 1.

Table 1. Calculated lattice constants a_0, bulk modulus B_0 first derivatives B_0' and gap energies ΔE for $Ga_xTl_{1-x}P$ compounds for x = 0.25, 0.50 or 0.75.

x	a_0 (Å)		B_0 (GPa)		B_0'		ΔE (eV)	Transition type
	Calculated	Literature	Calculated	Literature	Calculated	Literature		
0	6.3	6.12 [35]	42	46.75 [35]	4.67		-	
0.25	5.9		58.42		3.98		-	
0.5	5.78		59		4.27		0.37	Direct (Γ- Γ)
0.75	5.61		66.15		4.4		0.87	Direct (Γ- Γ)
1	5.43	5.43 [36]	91.15	91.15 [36]	3.6	3.6 [36]	1.5	Indirect (Γ-X)

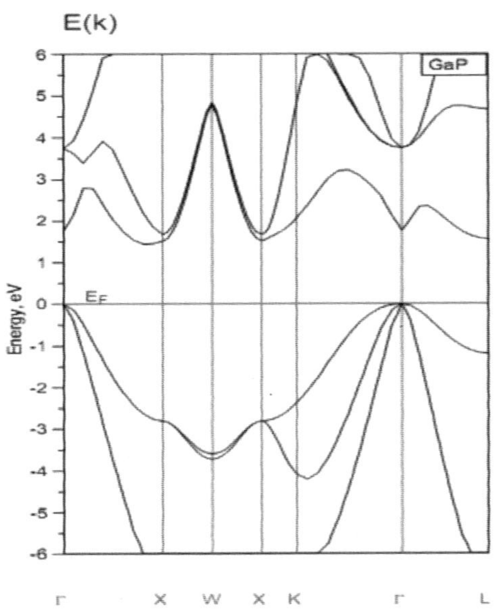

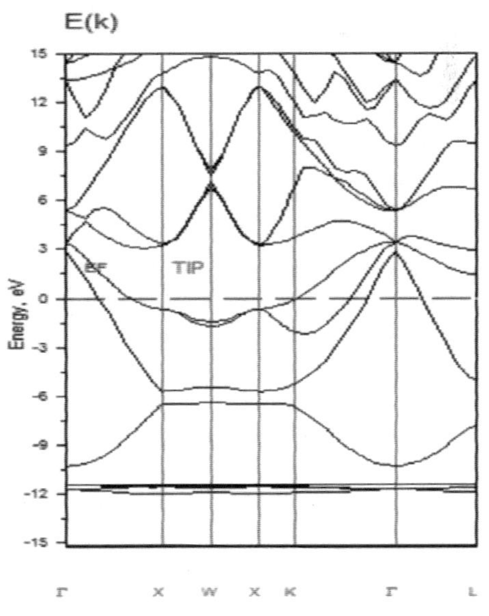

Figure 1. Band structure along the symmetry lines of the Brillouin zone at the equilibrium lattice constant for the (B3) structure for the binary alloys GaP (left graph) and TlP (right graph). The horizontal blue line shows the position of the Fermi level.

IOP Conf. Series: Journal of Physics: Conf. Series **1081** (2018) 012011 doi:10.1088/1742-6596/1081/1/012011

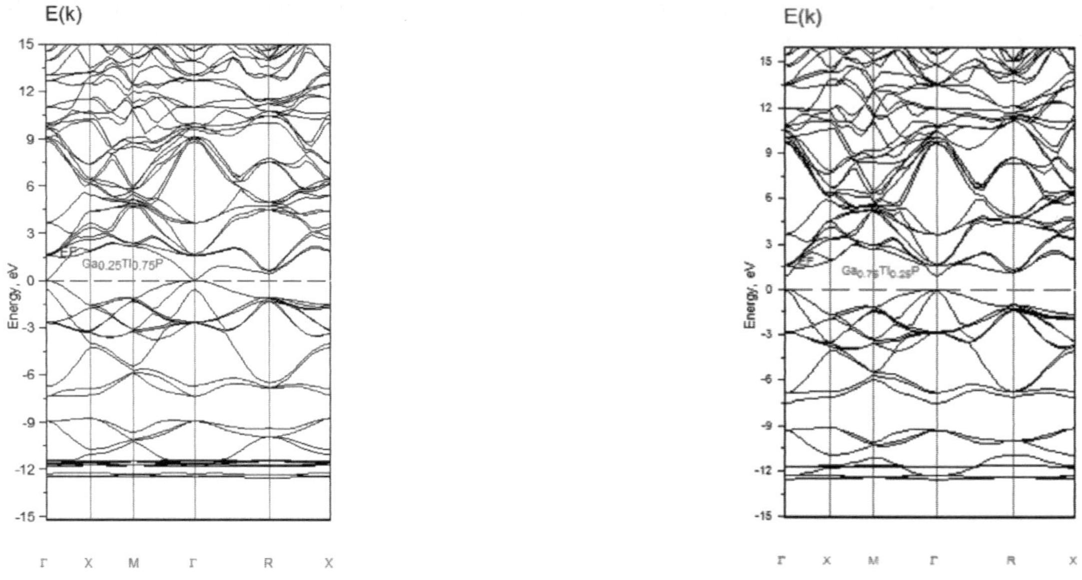

(a) (b) (c)

Figure 2. Band structures along the symmetry lines of the Brillouin zone at the equilibrium lattice constant for the (B3) structure for the $Ga_x Tl_{1-x} P$ ternary alloys (a) $Ga_{0.25} Tl_{0.75} P$, (b) $Ga_{0.5} Tl_{0.5} P$ and (c) $Ga_{0.75}Tl_{0.25}P$. The horizontal blue line shows the position of the Fermi level.

4. Conclusions

Full-potential linear muffin-tin method was used to investigate structural and electronic properties of the ternary alloys $Ga_x Tl_{1-x} P$. The lattice constants of these ternary alloys is correlated with increasing Tl concentration, but the bulk modulus is correlated inversely with increasing Tl concentration. Moreover, the gap energies of GaP and TlP binary compounds and their ternary alloys $Ga_x Tl_{1-x} P$ are decreasing rather linearly as Tl concentration increases. When compared with other published data, our calculated results gave an impetus for further application studies in optoelectronics, light-emitting diodes, laser diode, gas captors and solar cells.

References

[1] Madelung O, Landolt-BÖrnstein, Semiconductors 1982 Physics of Group IV Elements and III–V Compounds New Series, Berlin
[2] Zhang S B and Cohen M L 1987 *Phys. Rev.* B **35** 7604
[3] Causa M, Dovesi R and C. Roetti 1991 *Phys. Rev.* B **43** 11937
[4] Agrawal B K, Yadav P S, Kumar S and S. Agrawal 1995 *Phys. Rev.* B **52** 4896
[5] Paulus P, Fulde P and Stoll H 1996 *Phys. Rev.* B **54** 2556–
[6] Kalvoda S, Paulus B, Fulde P and Stoll H 1997 *Phys. Rev.* B **55** 4027
[7] Wang S Q and Ye H Q 2002 *J. Phys. Condens. Matter* **14** 9579
[8] Wang S Q and Ye H Q 2002 *Phys. Rev.* B **66** 235111
[9] Carrier P and Wei S H 2004 *Phys. Rev.* B **70** 035212
[10] Geller C B, Wolf W, Picozzi S, Continenza A, Asahi R, Mannstadt W, Freeman A J and Wimmer E 2001 *Appl. Phys. Lett.* **79** 368
[11] Mujica A, Robio A, Munoz A and Needs R J 2003 *Rev. Mod. Phys.* **75** 863
[12] Werking J D, Chang C R, Chang L D, Nguyen C and Kroemer H 1992 *IEEE Electron. Dev. Lett.* **13** 164

[13] Soderstrom J R, Chow D H and McGill T C 1989 *Appl. Phys. Lett.* **55** 1094
[14] Bouhafs B, Aourag H, Ferhat M and Certier M 1999 *J. Phys. Condens. Matter* **11** 5781
[15] Varshney D, Joshi G, Kaurav N and Singh R K 2009 *J. Phys. Chem. Solids* **70** 451
[16] Remediakis I N and Kaxiras E 1999 *Phys. Rev.* B **59** 5536
[17] Briki M, Abdelouhab M, Zaoui A and Ferhat M 2009 *Superlattice Microst.* **45** 80
[18] Wyckoff R W G 1986 Crystal Structures, second ed., Krieger, Malabar
[19] Madelung O 2004 Semiconductors: Data Handbook, Berlin
[20] Landolt-Bornstein 1982 Numerical Data and Functional Relationships in Science and Technology New Series, Berlin
[21] Kocinski P and Zbroszczyk M 1995 *Semicond. Sci. Technol.* **11** 1452
[22] Schroten E, Goossens A and Schoonman J 1998 *J. Appl. Phys.* **8** 1660
[23] Bouhafs B, Aourag H and Certier M 2000 *J. Phys. Condens. Matter* **12** 5655
[24] Zaoui A and El Haj Hassan F 2001 *J. Phys. Condens. Matter* **13** 253
[25] Touat D, Ferhat M and Zaoui A 2006 *J. Phys. Condens. Matter* **18** 3647
[26] Wentzcovitch R M and Cohen M L 1986 *J. Phys. C: Solid State Phys.* **19** 6791
[27] Ferhat M, Zaoui A, Certier M and Aourag H 1998 *Physica* B **252** 229
[28] Bouarissa N 2000 *Solid State Electron.* **44** 2193
[29] Ahmed R, Hashemifar S J and Akbarzadeh H 2008 *Physica B: Condens. Matter* **403** 1876
[30] Lin L, Woods G T and Callcott T A 2001 *Phys. Rev.* B **63** 235107
[31] Bouhemadou A, Khenata R, Kharoubi M, Seddik T, Reshak A H and Al-Douri Y 2009 *Comput. Mater. Sci.* **45** 474–79
[32] Arbouche O, Belgoumène B, Soudini B, Azzaz Y, Bendaoud H and Amara K 2010 *Comput. Mater. Sci.* **47** 685–92
[33] Vurgaftman I, Meyer J R and Ram-Mohan L R 2001 *J. Appl. Phys.* **89** 5815
[34] Adachi S 1994 GaAs and Related Materials, Bulk semiconducting and Superlattice Properties, Singapore
[35] Saidi N, Zaoui A and Ferhat M 2007 *Condens. Mater.* **19** 1062
[36] Merabet M, Benalia S and Rached D 2011 *Superlattice Microst.* **49** 139
[37] Savrasov S and Savrasov D 1992 *Phys. Rev.* B **46** 12181
[38] Savrasov S Y 1996 *Phys. Rev.* B **54** 6470
[39] Hohenberg P and Kohn W 1964 *Phys. Rev.* B **136** 864
[40] Kohn W and Sham L J 1965 *Phys. Rev.* A **140** 1133
[41] Perdew J P and Wang Y 1992 *Phys. Rev.* B **46** 12947
[42] Blochl P, Jepsen O and Andersen O K 1994 *Phys. Rev.* B **49** 16223
[43] Birch F 1978 *J. Geophys. Res.* **83** 1257

IOP Conf. Series: Journal of Physics: Conf. Series **1081** (2018) 012012 doi:10.1088/1742-6596/1081/1/012012

Sliding Mode Control for PMSG-based Wind Power System

Samir BELLARBI, D. SAHEB KOUSSA, A. DJOUDI
Centre de développement des énergies renouvelables CDER, BP. 62 Route de l'Observatoire Bouzareah, 16340, Alger, Algérie

E-mail : s.bellarbi@cder.dz

Abstract. Variable structure control has proven particularly suitable to variable speed wind energy conversion system control, being easy to implement by the already existing electronics. Its main drawback is an increase of mechanical stress due to chattering, which can however be alleviated by various methods. This paper proposes a sliding mode approach for tracking the energetic optimum of a variable speed permanent magnet synchronous generator PMSG. The sliding surface is systematically derived from imposing a desired reduced-order dynamics and allows the turbine operation more or less close to the optimal regimes characteristic, according to an imposed trade-off between the torque (control input) ripple and the optimum tracking. In this way, by torque controlling the generator, a multipurpose (energy-reliability) optimization is actually achieved.

1. Introduction

In recent decades, wind power becomes more interest as a source of electrical energy with minimal environmental impact. The variable wind speed turbine enables the operation over a wide speed range to extract the maximum power from wind. Recent trends indicate a move toward direct-drive turbine with permanent magnet synchronous generator PMSG. The PMSG have higher efficiency and power density as compared to wound rotor generators. Also, used to extract a maximum of energy by reducing mechanical stress by elimination of the speed multiplication (gear box), which improves system reliability and reduce maintenance costs [1].Several research studies on wind control were conducted. Thanks to them, the latest generation of wind turbines operates at variable speed, which offers a higher yield of energy compared to fixed-speed turbines [1] [2]. The rotational speed of turbine varies with changing the wind speed, which makes command of control schemes for electrical generator hard with wind turbine, since the wind speed varies.

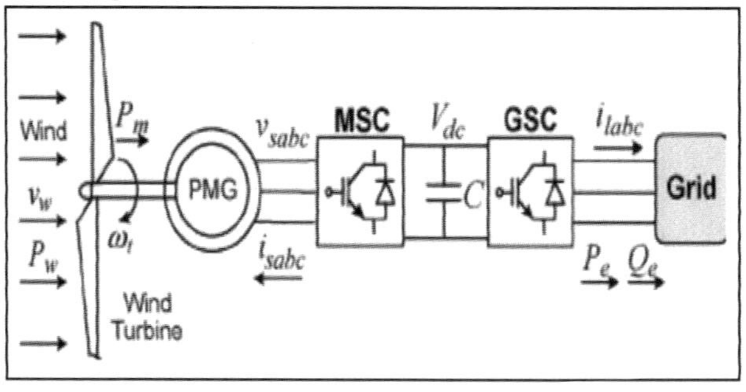

Figure 1. Wind energy conversion system based on PMSG [6].

Different studies of wind turbine were made, some use asynchronous motor, they have the advantage of having a low cost and easy maintenance. However, their speed control requires equipment in power electronics more expensive and more complex control [3] [4].
The purpose of wind energy systems is to maximize energy efficiency, and have a maximum power coefficient, and therefore extract maximum power from the wind speed. In This case, the MPPT control becomes important. To maintain, the operating point of maximum efficiency, different methods can be achieved. Optimum power/torque tracking is the most popular control strategy [5]. To realize this control, strategy conventional proportional and integral (PI) controller is usually used [5]. However, this strategy cannot achieve better performance. To solve this problem there are other control methods such as fuzzy logic, sliding mode control [6].

The sliding mode control is a control variable structure, which by its nature is a non-linear control. Proposed in the early 1950s, its success has been proven in control problems, since it is able to tackle system uncertainties and external disturbances with good robustness. This control is characterized by some advantages such as: high precision, rapid dynamic response, stability, simplicity of its design and its implementation [6]
Objective of this paper is propose a robust control to achieve better performance. This paper is organized as follows: describes The Wind Energy Conversion System (WECS). Sliding mode control is designed in develops control strategy and simulation results show the performance of the proposed approach and the conclusion is deducted in finally. The proposed strategy is compared with conventional PI controllers and confirmed superiorly in the MATLAB / Simulink environment.

2. Mathematical modelling of WECS

2.1. Turbine Model

The aerodynamic power P captured by the wind turbine is given by [5]:

$$P_{eol} = \frac{1}{2}.C_p(\beta,\lambda).\rho.S.V_V^3 \qquad (1)$$
$$\lambda = \frac{R\Omega}{V_V}$$

Where v is the wind, ρ is the air density, R is the rotor radius, C_p is the power coefficient and λ is the ratio of turbine blades tip speed to wind speed.

2.2.PMSG Model

Permanent magnet synchronous generator state model of the windings of the stator and the rotor according to the *d-q* axes are given by the following relations [6]:

$$x = [x_1(t) \;\; x_2(t)]^T = [i_d(t) \;\; i_q(t)]^T$$
$$u = [u_d \;\; u_q]^T \qquad (2)$$

$$\begin{cases} \dot{x} = \begin{bmatrix} -\frac{R}{L_d}x_1 + p\frac{L_q}{L_d}x_2\Omega_h \\ -\frac{R}{L_d}x_1 - p\frac{L_d x_1 - \Phi_m}{L_q}\Omega_h \end{bmatrix} + \begin{bmatrix} -\frac{1}{L_d} & 0 \\ 0 & -\frac{1}{L_d} \end{bmatrix}.u \\ y = \Gamma_G = p\Phi_m x_2 \end{cases} \qquad (3)$$

Where:
p: number of pole pairs.
R: stator resistance.
L_d, L_q: d and q inductances.
Φ_m : flux due to permanent magnets.
Γ_G: electromagnetic torque.
The electromagnetic torque given by:

$$\Gamma_G = -\frac{3}{2}.N_p.\Phi_f.i_{sq} \qquad (4)$$

IOP Conf. Series: Journal of Physics: Conf. Series **1081** (2018) 012012 doi:10.1088/1742-6596/1081/1/012012

The mechanical equation for the PMSG is expressed as:

$$J_{eq}.\frac{d\Omega_h}{dt} = \Gamma_h - \Gamma_G - f\Omega_h \qquad (5)$$

3. Sliding mode control

Sliding mode control is divided into three parts; first step is to select a sliding surface, which models the desired closed-loop performance in the state variable space. Then, we impose the condition of convergence, and then the command must be designed so that the system state trajectories are directed towards the sliding surface and stay on it [7] [8]. The principle operation is based on three steps:

- Selection of the sliding surface is based on the mathematical model of the PMSG.
- Convergence condition is obtained by the lyapunov function.
- Determine of the control strategy.

Considering the nonlinear system defined by the following form [9] [10]:

$$\dot{X} = g(X,t) + b(X,t)u(X,t) \qquad (6)$$

While: $X \in R^n, u \in R^m, g(X,t) \in R^n, b(X,t) \in R^{nxm}$ *g(X, t)* and *b(X, t)* are two non-linear functions, continuous, and limited.

The sliding surface is given by [8]:

$$S(X) = \left(\frac{d}{dt} + \delta\right)^{n-} e \qquad (7)$$

Where:

$$e = X^{ref} - X, X^{ref} = \left[x^{ref}, \dot{x}^{ref}, \ddot{x}^{ref}, \ldots\right]^T$$

e is the error, δ is a positive constant, X_{ref} is the reference signal, and X is the state variable.

Convergence condition: Lyapunov functions define the convergence condition:

$$S(X).\dot{S}(X) \leq 0 \qquad (8)$$

Control calculation: The control is given by this equation:

$$u = u^{eq} + u^n$$

Where u is the control signal, u^{eq} is the equivalent control, and u^n is the switching term which defined by the sign of the sliding surface multiplied by a positive constant:

$$u^n = k.sat(S(X)) \qquad (9)$$

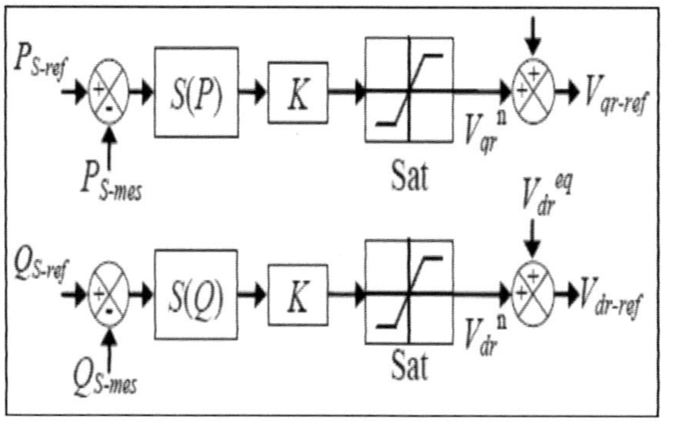

Figure 2. Sliding mode controller diagram [7].

3.1. *Sliding control for wind speed turbine*

The sliding surface and its derivative for the mechanical speed generator are given by:

$$S_\Omega = \Omega_{hopt} - \Omega_h$$
$$\frac{dS_\Omega}{dt} = \frac{d\Omega_{hopt}}{dt} - \frac{d\Omega_h}{dt} \quad (10)$$

Substituting (5) in (10), the new derivative of sliding surface is shown in (11).

$$\frac{dS_\Omega}{dt} = \frac{d\Omega_{hopt}}{dt} - \frac{1}{J_{eq}}(\Gamma_h - \Gamma_G - f\Omega_h) \quad (11)$$

The Lyapunov function is defined by:

$$v = \frac{1}{2} S_\Omega^2 \quad (12)$$

According to condition (8) we obtain:

$$\frac{dv}{dt} = S_\Omega . \frac{dS_\Omega}{dt} \le 0 \quad (13)$$

In order to obtain the control strategy for Ω_h, the electromagnetic torque which is the variable command is composed of two components:

$$\Gamma_h = \Gamma_{h.eq} - \Gamma_{h.n} \quad (14)$$

Where, $\Gamma_{h.eq}$ is the equivalent command enables the pursuit of the trajectory on sliding surface, and it is deduced from $\dot{S}_\Omega = 0$. The other component Γ_hn is the discontinuous control which is responsible of the attractiveness of the variable command around the sliding surface, it is calculated by:

$$\Gamma_{h.n} = \eta_\Omega . sat\left(\frac{S_\Omega}{\varepsilon}\right) \quad (15)$$

η_Ω is the gain of sliding mode controller, the saturation function of is given by (16), and is the width of boundary layer of the sliding surface.

$$sat\left(\frac{S_\Omega}{\varepsilon}\right) = \begin{cases} sign\left(\frac{S_\Omega}{\varepsilon}\right); |S_\Omega| \rangle \varepsilon \\ \frac{S_\Omega}{\varepsilon}; |S_\Omega| \le \varepsilon \end{cases} \quad (16)$$

By solving $\dot{S}_\Omega = 0$, it found that $\Omega_{opt} = \Omega_h$. Then the equivalent command of electromagnetic torque is given by:

$$\Gamma_{h.eq} = \Gamma_h - f\Omega_h - J_{eq}.\dot{\Omega}_{h\_opt} \quad (17)$$

Then, the electromagnetic torque command is expressed as (18) ; substituting (4) obtain the reference q-axis current in (18):

$$\Gamma_G = \Gamma_h - f\Omega_h - J_{eq}.\dot{\Omega}_{h\_opt} + \eta_\Omega.sat\left(\frac{S_\Omega}{\varepsilon}\right) \quad (18)$$

$$i_{sqr} = \frac{-2}{3.N_p.\Phi_f}\left[\Gamma_h - f\Omega_h - J_{eq}.\dot{\Omega}_{hopt} + \eta_\Omega.sat\left(\frac{S_\Omega}{\varepsilon}\right)\right] \quad (19)$$

The sliding surfaces for the *dq*-axis current control are:

$$\begin{cases} S_d = i_{sdr} - i_{sd} \\ S_q = i_{sqr} - i_{sq} \end{cases} \quad (20)$$

By (10) and (19), we obtain:

$$v_{sdeq} = \Omega_h.L_q.i_{sq} - R_s.i_{sd}$$
$$v_{sqeq} = -L_q.\frac{di_{sqr}}{dt} - R_s.i_{sq} - \Omega_h.L_d.i_{sd} + \Omega_h.\Phi_f \quad (21)$$

Then, v_{sdh} and v_{sqh} are given by the same method as shown by:

$$v_{sd\,h} = \Omega_h.L_q.i_{sq} - R_s.i_{sd} + \eta_d.sat\left(\frac{S_d}{\varepsilon}\right)$$

$$v_{sq\,h} = -L_q.\frac{di_{sqr}}{dt} - R_s.i_{sq} - \Omega_h.L_d.i_{sd} + \Omega_h.\Phi_f + \eta_q.sat\left(\frac{S_q}{\varepsilon}\right) \qquad (22)$$

3.2.. Sliding Control for Grid Side Converter

The following equations for Grid Side Converter are:

$$V_{id} = (R_f + s.L_f)i_{gd} - \Omega_g.L_f.i_{gq} + v_{gd}$$

$$V_{iq} = (R_f + s.L_f)i_{gq} + \Omega_g.L_f.i_{gd}$$

$$C\frac{dV_{dc}}{dt} = i_g - \frac{3}{2}\left(\frac{V_{gd}}{V_{dc}}.i_{gd}\right) \qquad (23)$$

Where, V_{idq} is dq-axis inverter voltage components, i_{gdq}; V_{gdq} are the dq-axis grid current and voltage components respectively, V_{dc}; i_{dc} are the dc-bus voltage and current, i_g is the output current of the generator side converter, and R_f,L_f are resistance and inductance of R_L filter, Ω_g is the grid frequency. The active and reactive power is expressed by:

$$\begin{cases} P=\frac{3}{2}(V_{gd}.i_{gd}+V_{gq}.i_{gq})=\frac{3}{2}V_{gd}.i_{gd} \\ Q=\frac{3}{2}(V_{gq}.i_{gd}-V_{gd}.i_{gq})=-\frac{3}{2}V_{gd}.i_{gd} \end{cases} \qquad (24)$$

$$i_{dqr} = \frac{-2Q_r}{3.V_{gd}} \qquad (25)$$

The sliding surfaces for the DC-bus voltage and dq-axis grid currents are expressed by:

$$S_{dc} = V_{dcr} - V_{dc}$$
$$S_{gd} = i_{gdr} - i_{gd} \qquad (26)$$
$$S_{gq} = i_{gqr} - i_{gq}$$

The convergence condition is expressed by:

$$\dot{S}_{dc} = \dot{V}_{dcr} - \dot{V}_{dc}$$
$$\dot{S}_{gd} = \dot{i}_{gdr} - \dot{i}_{gd} \qquad (27)$$
$$\dot{S}_{gq} = \dot{i}_{gqr} - \dot{i}_{gq}$$

The expressions of variable command for different controllers are expressed by:

$$i_{gd\,h} = \frac{2}{3}.\frac{i_g.V_{dc}}{V_{gd}} + \eta_{dc}.sat\left(\frac{S_{dc}}{\varepsilon}\right)$$

$$V_{id\,h} = L_f.\frac{di_{gdr}}{dt} + R_f.i_{gdr} - \Omega_g.L_f.i_{gqr} + V_{gd} + \eta_{gd}.sat\left(\frac{S_{gd}}{\varepsilon}\right) \qquad (28)$$

$$V_{iq\,h} = R_f.i_{gqr} - \Omega_g.L_f.i_{gdr} + \eta_{gq}.sat\left(\frac{S_{gq}}{\varepsilon}\right)$$

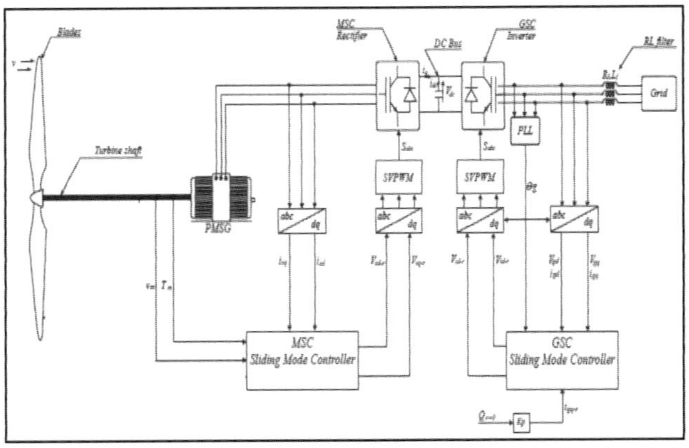

Figure 3: Control scheme of the WECS [9].

Figure 4 shows that the operating point is attracted to the new sliding surface and then evolves in sliding mode to the optimal one.

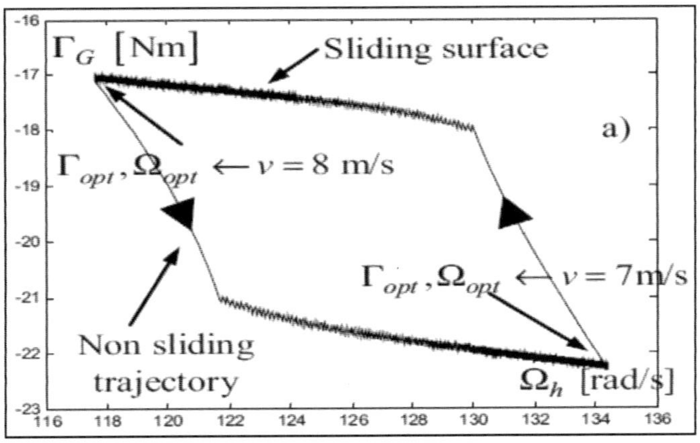

Figure 4: WECS state space evolution by SMC [10].

4. Results and Discussion

MATLAB/Simulink numerical simulations have been performed in order to assess the sliding mode optimization approach detailed above. Simulations detailed have been done for 180 seconds time horizon, using a pseudo-random sequence of wind speed. Then, we compare the performance of the SMC controllers with the PI controller [11].

Initially, a wind speed profile with a maximum of 11 m/s and a minimum of 7 m/s was applied to the PMSG WECS in order to see the performance of the proposed controller.

The wind speed, power coefficient, wind power, active power, reactive power, stator voltage and stator current are shown respectively in figures 5, 6, 7, 8, 9 and 10 where a good performance was observed.

Power coefficient value it's equal to 0.47 for SMC controller where it was found equal to 0.44 in PI controller (C_{pmax}= 0.48), which shows good performance of SMC controller.

The active power P and reactive power Q are regulated. In figure10, it can be seen that the current and voltage are in phase, which is established by controlling the power factor. It can therefore be concluded from the results of simulation that the SMC controller has good performance of tracking power points. The method also has good application potential in other types of wind energy conversion system.

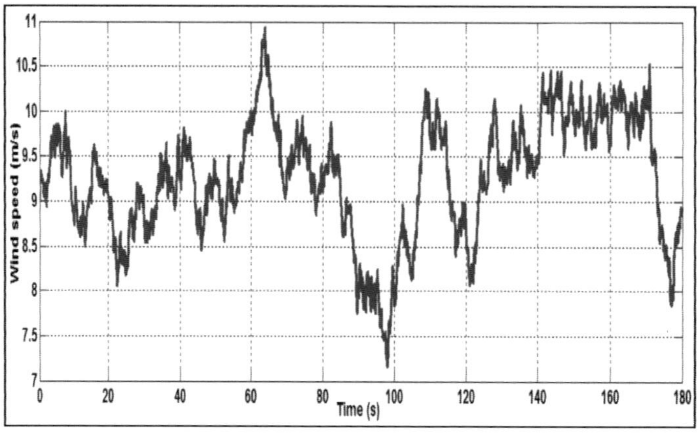

Figure 5. Wind speed profile.

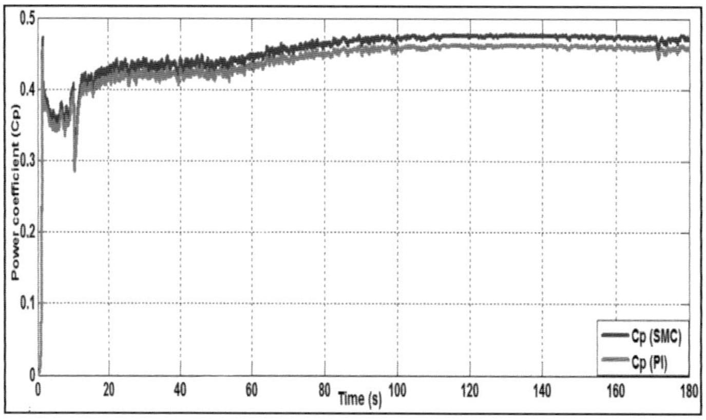

Figure 6. power coefficient profile

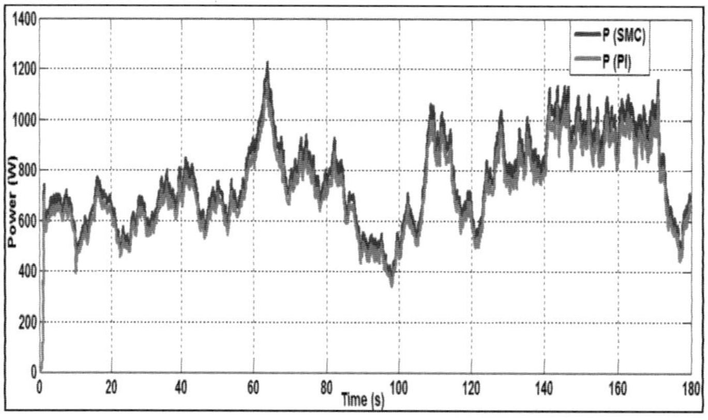

Figure 7. Wind turbine power profile.

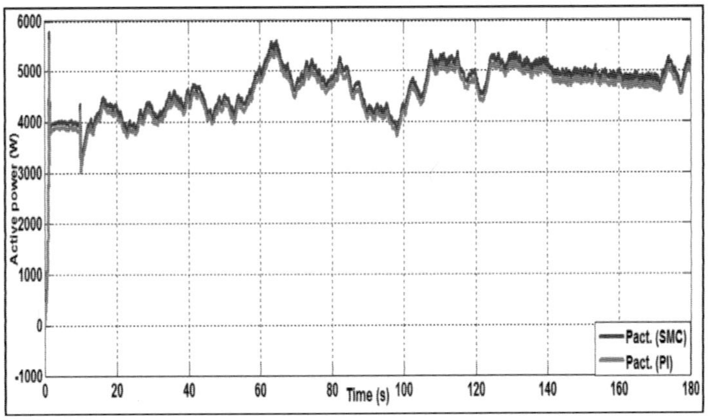

Figure 8. Acitve power profile.

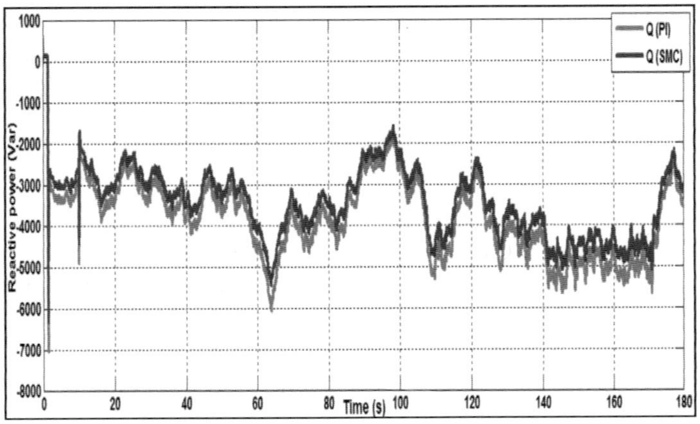

Figure 9. Reacitve power profile.

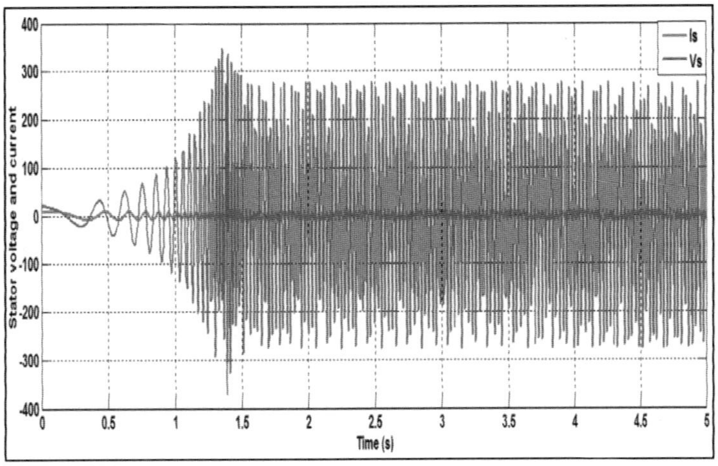

Figure 10. Stator voltage and current by phase.

5. Conclusion

The sliding mode control law presented in this paper aims at maximizing the wind harvested power, while limiting the electromagnetic torque variations. Provided that ensuring the optimal tip speed ratio implies strong torque variations, the choice of an appropriate sliding surface is not a trivial task. Such a surface, having a non-empty intersection with the optimal regimes characteristic, variant with the wind speed and depending on a desired energy reliability trade off, has been systematically found.

The possibility of driving the operating point in a conveniently sized neighborhood of the optimal regimes characteristic is thus ensured by imposing some desired reduced order dynamics, implicitly allowing the generator torque variations limitation in the high frequency range. The drive train mechanical fatigue induced by the generator control is alleviated with positive influence on the WECS overall reliability. The possibility of adjusting the trade-off coefficient, confers flexibility to the WECS, so that the wind energy conversion efficiency be significantly increased when the particular conditions of the site allow it.

References

[1]Q. Huang and Z. Pan, “Sensor less control of permanent magnet synchronous generator in direct-drive wind power system,” in 2011 International Conference on Electrical Machines and Systems (ICEMS), 2011, pp. 1–5.

[2] K. Z. Østergaard, J. Stoustr up, and P. Brath, “Linear parameter varying control of wind turbines covering both partial load and full load conditions,” Int. J. Robust Nonlinear Contr., vol. 19, pp. 92–116, 2009.

[3] S. Bellarbi, D. Saheb Koussa, "Fuzzy Robust Control of DFIG with parameter uncertainties", "3rd Euro-Mediterranean Conference on Materials and Renewable Energies" (EMCMRE-3), Marrakech, Morocco 2015.

[4] S. Bellarbi, D. Saheb Koussa, A. Djoudi, “Power Comparison between DFIG and SCIG for Wind Energy System ,” IEEE, International Renewable and Sustainable Energy Conference - IRSEC'16 - 14-17 October 2016 Marrakech (Maroc).

[5]. S. Bellarbi, D. Saheb Koussa, A. Rennane, "Power Control of a Wind Energy Conversion System", IEEE, 6th International Renewable Energy Congress, page 1-4, March 2015.

[6] Y. Errami, M. Ouassaid, M, “Sliding Mode Control Scheme of Variable Speed Wind Energy Conversion System Based on the PMSG for Utility Network Connection,” in Advances and Applications in Sliding Mode Control systems, ser. Studies in Computational Intelligence. Springer International Publishing, 2015, no. 576, pp. 167–200.

[7] Boukezzar, Boubekeur, and Mohamed M'Saad, "Robust sliding mode control of a DFIG variable speed wind turbine for power production optimization." Control and Automation, 2008 16th Mediterranean Conference on. IEEE, 2008.

[8] Slotine JJE, Li W, “Applied Nonlinear Control,” Englewood Cliffs, NJ, USA: Prentice Hall, 1998.

[9] E.G. Shehata, “Sliding mode direct power control of RSC for DFIGs driven by variable speed wind turbines,” Alexandria Engineering Journal, Elsevier, Volume 54, Issue 4, December 2015, Pages 10671075.

[10] Zhang Yan and V. Utkin, “Sliding mode observers for electric machines an overview,” vol. 3. IEEE, 2002, pp. 1842–1847.

[11] Y. Kim, Y. Chung, “Tuning of the PI Controller Parameters of a PMSG Wind Turbine to Improve Control Performance under Various Wind Speeds,” Energies 2015, 8(2), 1406-1425; doi:10.3390/en8021406.

Annex

PMSG parameters are:

Rated power: P_n= 1.4MW; V = 690 V and f =50Hz, Stator resistance: Rs = 0,821$m\Omega$; dq-axis inductances: L_{dq} = 1,573mH; Number of pole pair: N_p = 26; Rated mechanical torque: T_{nom}=848.826 KN.m; PM flux : Φ_f= 5.8264Wb